Lecture Notes in Computer Science 4268

Commenced Publication in 1973
Founding and Former Series Editors:
Gerhard Goos, Juris Hartmanis, and Jan van Leeuwen

Gerard Parr   David Malone
Mícheál Ó Foghlú (Eds.)

Autonomic Principles of IP Operations and Management

6th IEEE International Workshop
on IP Operations and Management, IPOM 2006
Dublin, Ireland, October 23-25, 2006
Proceedings

 Springer

Volume Editors

Gerard Parr
University of Ulster
School of Computing and Information Engineering
Coleraine Campus, Cromore Road, Coleraine, BT52 1SA, Northern Ireland
E-mail: gp.parr@ulster.ac.uk

David Malone
National University of Ireland, Maynooth
Hamilton Institute
Maynooth, Kildare, Ireland
E-mail: david.malone@nuim.ie

Mícheál Ó Foghlú
Waterford Institute of Technology
Telecommunications Software & Systems Group
Cork Road, Waterford, Ireland
E-mail: mofoghlu@tssg.org

Library of Congress Control Number: 2006934472

CR Subject Classification (1998): C.2, D.4.4, D.2, H.3.5, H.4, K.6.4

LNCS Sublibrary: SL 5 – Computer Communication Networks and
Telecommunications

ISSN   0302-9743
ISBN-10   3-540-47701-2 Springer Berlin Heidelberg New York
ISBN-13   978-3-540-47701-3 Springer Berlin Heidelberg New York

Springer is a part of Springer Science+Business Media

springer.com

© Springer-Verlag Berlin Heidelberg 2006
Printed in Germany

Typesetting: Camera-ready by author, data conversion by Scientific Publishing Services, Chennai, India
Printed on acid-free paper   SPIN: 11908852   06/3142   5 4 3 2 1 0

Preface

This volume presents the proceedings of the *6th IEEE International Workshop on IP Operations and Management (IPOM 2006)*, which was held as part of Manweek 2006 in Dublin, Ireland from October 23rd to 25th, 2006. In line with its reputation as one ofthe pre-eminent venues for the discussion and debate of advances of management of IP networks and services, the 2006 iteration of IPOM brought together an international audience of researchers and practitioners from both industry and academia. The overall theme of Manweek 2006 was "Autonomic Component and System Management",with IPOM taking this to be the application of autonomic principles to the IP operations, administration, maintenance and provisioning (OAM&P) domain.

IPOM 2006 is more relevant than ever to the emerging communications infrastructure that is increasingly focused on "convergence" of networks and services. Although arguably over-hyped, there is a fundamental truth to this convergence story, and this is based on the fact that the TCP/IP protocol suite (IPv4 and IPv6) has become the common denominator for a plethora of such converged services. One good example in the period between IPOM 2005 and IPOM 2006 has been the large scale deployment of consumer VoIP, linked to the success of Skype and alternatives including SIP-based approaches. In many countries VoIP is driving broadband deployment for SMEs where real costs savings can be accrued, especially for companies with remote staff in the field. Many operators are now deploying Quality of Service (QoS) schemes to manage this VoIP (and other premium) traffic. This brings these issues from the research laboratory into the operations and management domain.

Being a relatively pragmatic workshop IPOM 2006 is focused on issues that matter to those managing such IP networks and services, both enterprise networks and telecommunications operators' networks. These issues include the complexity of interoperability between networks and service providers, the performance versus costs in operating IP-based networks, and the OAM&P challenges in next generation networks(NGNs) and related seamless service provision. Of particular interest in the telecommunications sector are issues related to Fixed-Mobile Convergence and the emerging IP Multimedia System (IMS). These issues were reflected in the issued call for papers.

In response to the IPOM 2006 call for papers a total of 45 paper submissions were received from the research community. Of these, 39 were full papers and 6 were short papers. After a comprehensive review process carried out by the technical programme committee and additional subject area experts all submissions were ranked based on review scores and the co-chair's view on their contribution and relevance to the conference scope. All submissions received at least 3 reviews, with most receiving 4. After lengthy discussions it was decided to accept 18 of the 39 submitted full papers (40% acceptance rate of the total submissions) and 4 short papers. These papers present novel and interesting contributions in topics ranging from OSPF weightings in intradomain QoS, to large scale topology discovery. We believe that, taken together, these papers provide a provocative insight into the current state of the art in IP operations and management.

There are many people whose hard work and commitment were essential to the success of IPOM 2006. Foremost amongst these are the researchers who submitted papers to the conference. The overall quality of submissions this year was high and we regret that many high quality papers had to be rejected. We would like to express out gratitude to both the IPOM steering committee and the technical committee, for their advice and support through all the stages of the conference preparation. We thank all paper reviewers, in particular those outside the technical programme committee, for their uniformly thorough, fair and helpful reviews. We thank the IEEE for their continued support and sponsorship of IPOM.

Most of the time-consuming practical and logistical organisation tasks for the conference were handled by the members of the Manweek Organisation Committee – this made our jobs significantly easier, and for that we are very grateful. Finally, we wish to acknowledge the financial support of both Science Foundation Ireland and the Manweek corporate sponsors, whose contributions were hugely instrumental in helping us run what we hope was a stimulating, rewarding and, most importantly, an enjoyable conference for all its participants.

October 2006

Gerard Parr
David Malone
Mícheál Ó Foghlú

IPOM 2006 Organisation

Technical Programme Committee Co-chairs

Gerard Parr   University of Ulster, UK
David Malone   NUI Maynooth, Ireland
Mícheál Ó Foghlú   Waterford Institute of Technology, Ireland

Steering Committee

Tom Chen   Southern Methodist University, USA
Petre Dini   Cisco Systems, USA
Andrzej Jajszczyk   AGH University of Science and Technology, Poland
G.-S. Kuo   NCCU, Republic of China
Deep Medhi   University of Missouri-Kansas City, USA
Curtis Siller   IEEE ComSoc, USA

Organisation Co-chairs

Brendan Jennings   Waterford Institute of Technology, Ireland
Sven van der Meer   Waterford Institute of Technology, Ireland

Publication Chair

Tom Pfeifer   Waterford Institute of Technology, Ireland

Publicity Co-chairs

Sasitharan Balasubramaniam   Waterford Institute of Technology, Ireland
John Murphy   University College Dublin, Ireland

Treasurer

Mícheál Ó Foghlú   Waterford Institute of Technology, Ireland

Local Arrangements

Miguel Ponce de León   Waterford Institute of Technology, Ireland
Dave Lewis   Trinity College Dublin, Ireland
Dirk Pesch   Cork Institute of Technology, Ireland

Gabriel-Miro Muntean	Dublin City University, Ireland
Seán Murphy	University College Dublin, Ireland
Rob Brennan	Ericsson, Ireland

Manweek 2006 General Co-chairs

William Donnelly	Waterford Institute of Technology, Ireland
John Strassner	Motorola Labs, USA

Manweek 2006 Advisors

Raouf Boutaba	University of Waterloo, Canada
Joan Serrat	Universitat Politècnica de Catalunya, Spain

IPOM 2006 Technical Programme Committee

Nader Azarmi	BT Group Research, UK
John-Luc Bakker	Telcordia, USA
Saleem Bhatti	University of St Andrews, UK
Marcus Brunner	NEC Europe, Germany
Baek-Young Choi	University of Missouri-Kansas City, USA
Alexander Clemm	Cisco Systems, USA
Haitham Cruickshank	University of Surrey, UK
Laurie Cuthbert	Queen Mary University of London, UK
Timothy Gonsalves	IIT Madras, India
Abdelhakim Hafid	University of Montreal, Canada
Steve Hailes	University College London, UK
Masum Hasan	Cisco Systems, USA
David Hutchison	Lancaster University, UK
Wolfgang Kellerer	DoCoMo Eurolabs, Germany
G.S. Kuo	NCCU, Republic of China
Edmundo Madeira	UNICAMP, Brazil
Thomas Magedanz	Fraunhofer FOKUS, Germany
Manu Malek	Stevens Institute of Technology, USA
Dave Maltz	Microsoft, USA
Deep Medhi	University of Missouri-Kansas City, USA
Maurizio Molina	DANTE, UK
Donal O'Mahony	Trinity College Dublin, Ireland
Michal Pioro	Warsaw University of Technology, Poland
Caterina Scoglio	Kansas State University, USA
Bryan Scotney	University of Ulster, UK
Stephan Steglich	Fraunhofer FOKUS, Germany
Martin Stiemerling	NEC, Germany

John Strassner   Motorola Labs, USA
Vincent Wade   Trinity College Dublin, Ireland

IPOM 2006 Additional Paper Reviewers

Sasitharan Balasubramaniam   Waterford Institute of Technology, Ireland
Keara Barrett   Waterford Institute of Technology, Ireland
Prakash Bettadapur   Cisco Systems, USA
Dmitri Botvich   TSSG, Ireland
Ray Carroll   Waterford Institute of Technology, Ireland
Peter Clifford   NUI Maynooth, Ireland
Steven Davy   Waterford Institute of Technology, Ireland
Petre Dini   Cisco, USA
William Fitzgerald   Waterford Institute of Technology, Ireland
Paulo Freitas   UFSC, Brazil
Luciano Gaspary   UFRGS, Brazil
Celio Guimaraes   UNICAMP, Brazil
Paul Malone   Waterford Institute of Technology, Ireland
Joberto Martins   UNIFACS - Universidade Salvador, Brazil
Jimmy McGibney   Waterford Institute of Technology, Ireland
Sven van der Meer   Waterford Institute of Technology, Ireland
Niall Murphy   Amazon.COM Network Engineering, Ireland
Seán Murphy   University College Dublin, Ireland
Venet Osmani   Waterford Institute of Technology, Ireland
Tom Pfeifer   TSSG, Ireland
Miguel Ponce de Leon   TSSG, Ireland
John Ronan   Waterford Institute of Technology, Ireland
José Augusto Suruagy Monteiro   UNIFACS, Brazil
Fábio Verdi   Unicamp, Brazil
Rolf Winter   NEC Europe, Germany

Table of Contents

1. Modeling and Planning

2. Quality of Service Routing

3. Quality of Service Issues

4. Management and Configuration

5. Autonomics and Security

6. Topology

7. Short Papers

Traffic Modeling and Classification Using Packet Train Length and Packet Train Size

Dinil Mon Divakaran, Hema A. Murthy, and Timothy A. Gonsalves

Department of Computer Science and Engineering
Indian Institute of Technology, Madras
Chennai - 600036
{dinil, hema, tag}@tenet.res.in

Abstract. Traffic modeling and classification finds importance in many areas such as bandwidth management, traffic analysis, traffic prediction, network planning, Quality of Service provisioning and anomalous traffic detection. Network traffic exhibits some statistically invariant properties. Earlier works show that it is possible to identify traffic based on its statistical characteristics. In this paper, an attempt is made to identify the statistically invariant properties of different traffic classes using multiple parameters, namely packet train length and packet train size. Models generated using these parameters are found to be highly accurate in classifying different traffic classes. The parameters are also useful in revealing different classes of services within different traffic classes.

1 Introduction

The phenomenal expansion of Internet has seen a rapid growth in the number and variety of applications. Many of such applications turn out to be bandwidth-hungry or delay-sensitive; and require, or at least benefit from specific service classes that prioritize packets in the Internet. Internet Service Providers and network operators need to classify traffic data within their network and evaluate their absolute and relative importance and subsequently create traffic policies to be enforced in the Internet routers [1]. Network administrators need to know the different classes or types of traffic that flow through their network so as to manage the network efficiently. Accurate traffic classification is essential for provisioning and bandwidth management. Floyd and Paxson [2] pointed out that it is important to capture the invariants of traffic to cope with the constantly changing nature of Internet traffic.

Conventional methods for traffic classification use the packet header information to find out the ports used for communication. Well known ports are supposed to be used by specific application protocols (e.g. port 80 is usually used by HTTP). But this method has become less and less accurate as more and more emerging applications use well know ports for relaying traffic, e.g. tunneling over HTTP port is very common [3,4]. Owing to the increasing traffic and application protocols, techniques based on statistical modeling are gaining importance. Recent works have made efforts to increase the accuracy in traffic classification [5,6,7,8].

G. Parr, D. Malone, and M. Ó Foghlú (Eds.): IPOM 2006, LNCS 4268, pp. 1–12, 2006.

In traffic modeling, parameters are extracted from packet headers. There are different parameters such as packet length, packet inter-arrival time, flow duration, packet train inter-arrival time, packet train length, packet train size etc that can be considered for modeling of traffic. While parameters can be modeled separately [9], we focus on modeling the traffic classes using multiple parameters.

In [10], it was shown that traffic characteristics are generally multimodal in nature. For example, the total number of packets transferred during mail transfer vary for small text messages to large picture attachments. To capture the multimodal characteristic of traffic, we employ clustering techniques based on Vector Quantization (VQ) [11] and Gaussian Mixture Models (GMM) [12]. The models obtained using these techniques are later used for classification of a given data set (not used during training) into one of the different traffic types.

The rest of the paper is organized as follows. In Sect.3 we explain the experimental setup. Modeling and classification using VQ is explained in Sect.4. Section 5 details modeling using GMM and Bayesian classification using the Gaussian mixtures obtained. Evaluation and verification of models follow in Sect.6. We conclude in Sect.7.

2   Related Work

System administrators have long being using port based classification method to identify different traffic classes flowing in the network. Tools such as tcpdump [13] read the header information to find the source and destination ports. Each server port associates itself with an application as per the IANA (Internet Assigned Numbers Authority) [14], which maintains a mapping of the server ports to application types. Such a method does not provide good accuracy for a number of reasons. For instance, with the proliferation of applications, not every application is registered with IANA. Users behind a firewall that permits packets to only a few ports, usually relay traffic through well known ports (eg. SSH over HTTP). Similarly, non-privileged users run HTTP servers on ports other than 80. Even for applications defined with IANA, some ports are used by different applications with entirely different QoS requirements (for example, SSH and SCP use same port 22).

Statistical traffic classification is an alternative to the less accurate port based classification. Past research works have focused on characterizing particular traffic classes. In [15], joint distribution of flow duration and number of bytes were used to identify DNS traffic. Paxson [16] examined the distribution of flow bytes and packets for a number of different applications. Roughan et al. used LDA (Linear Discriminant Analysis) and QDA (Qualitative Discriminant Analysis) to classify traffic into different classes of services for QoS implementation [5]. In [6], authors describe a method for visualisation of the attribute statistics that aids in recognizing cluster types. The method uses EM (Expectation-Maximistaion) for probabilistic clustering with parameters such as packet size and packet inter-arrival time. In [8], the authors explore Bayesian classifier using a number of per flow discriminators to classify network traffic.

The multimodal nature of traffic was highlighted in [9] using packet size as a promising parameter for traffic characterisation. We extend this work and use flow related information for modeling traffic.

3   Traffic Modeling

In this section, we detail the setup for experimentation in terms of the parameters used for identifying the various traffic classes considered for this work.

3.1   Parameters

The goodness of these models largely depends on the parameters that are used for modeling. The parameters used here for modeling and classification of traffic classes are *packet train length* and *packet train size*. The concept of packet train was introduced by Raj Jain and Shawn A. Routhier [17]. The two ends of a packet train are identified as two nodes in a network. As defined in [17], a packet train is essentially the flow of packets between two nodes in a network, where each packet forms the car of the train. Here, we modify the definition of a packet train to be the flow between two sockets, which is appropriately identified by the quadruple *(source host, source port, destination host, destination port)*. Each end of such a packet train is identified by the node name and port number. Packet train length is then defined as the number of packets within a train; and packet train size is the sum of the sizes of all the packets that form a packet train. The advantage of using these parameters (as will be discussed later), is that the models generated using these parameters give information regarding the class of service within an application type (say HTTP), apart from classifying traffic.

Since we consider two parameters for modeling and testing, it is important to ensure that one parameter doesn't overshadow the other parameter. Appropriate normalization of parameters is therefore performed.

3.2   Traffic Classes

Five commonly used application protocols are selected for modeling and testing. These are HTTP, SMTP, DNS, SSH and POP3 which use ports 80, 25, 53, 22 and 110 respectively. DNS uses UDP, whereas HTTP, SMTP, SSH and POP3 use TCP as the transport layer protocol. The terms *traffic types* and *traffic classes* refer to these application protocols in general.

This work looks only at one UDP based traffic class, namely DNS traffic. Since all other traffic classes considered are TCP based, and hence connection oriented, they inherently have the packet train property. But, it should be noted that almost all UDP based applications can be viewed as a connection oriented traffic and therefore can be represented using packet trains. For example, a video conferencing tool running on top of UDP can be identified uniquely by *src host*, *src port*, *dst host* and *dst port*, and all such packets can be considered as part of a connection. To distinguish different connections of the same applications, the time between packets can be used; as the time between consecutive packets of one run of the application will be much less as compared to time between consecutive packets of different runs of the application. Hence, using all these information, packet train parameters can be extracted for the traffic generated by a video conferencing application, or in general, for traffic generated by almost any UDP based application.

3.3   Data Representation

Throughout the paper, data or packet trains are represented as a set of N vectors, $X = \{x_1, x_2, ..., x_N\}$ [1]. Each input vector has a dimension of two corresponding to the two parameters used in modeling.

4   Classification Using Vector Quantization

VQ is a very popular approximate method in the class of clustering algorithms that simplifies computations and accelerates convergence. VQ partitions d-dimensional vectors in the vector space, $\mathbf{R}^d$, into finite sets of vectors based on the *nearest-neighbour* criterion [18]. Such sets, called clusters, represent separate regions in the vector space. A vector, $\mathbf{x} \in \mathbf{R}^d$, belongs to cluster C_i, if

$$\left\| \mathbf{x} - \boldsymbol{\mu_i} \right\| < \left\| \mathbf{x} - \boldsymbol{\mu_j} \right\| \quad \text{for all } j \neq i \ . \tag{1}$$

where $\boldsymbol{\mu_i}$ is the mean vector of the cluster, C_i. This equation states that a vector belongs to the nearest cluster. If there are two or more clusters to which the distance from the vector is minimum, one among them is chosen randomly. The clusters partition the vector space such that

$$\bigcup_{i=1}^{k} C_i = \mathbf{R}^d \quad and \quad \bigcap_{i=1}^{k} C_i = \phi \ . \tag{2}$$

where k is the number of clusters.

4.1   Training

During training, we use VQ to partition the vector space, where the first dimension of each vector is *packet train length* and second dimension is *packet train size*. The algorithm used is as follows

1. Initialize the mean (vector) of each cluster by randomly selecting a vector from the given set of vectors, $\mathbf{X}$, such that no two clusters have the same mean.
2. Until the mean of each cluster converges
   - Classify each vector into one of the clusters using (1).
   - Recompute the mean of each cluster.

Using the above algorithm, the vectors in the given data set, $\mathbf{X}$, are classified into clusters. The models thus generated for each traffic type is used for testing.

---

[1] Boldface is used to denote vectors and matrices.

4.2   Testing

In the testing phase, we have to find the traffic type that is *nearest* to the given data set, $\mathbf{X}$. For this, we calculate the distance between the ith vector, $\mathbf{x_i}$, and the nearest cluster (best cluster) of a particular traffic type, s, using the Euclidean distance function

$$dist(\mathbf{x}_i, s) = \min_j \sqrt{\sum_{l=1}^{d} \left[\mathbf{x_{i_l}} - \boldsymbol{\mu_{j_l}}\right]^2} \qquad 1 \le j \le n(s) . \tag{3}$$

where $n(s)$ is the number of clusters in the traffic type s, and d is the dimension of the vectors. It should be noted that, since the clusters depend on the traffic type, the mean vector, $\boldsymbol{\mu_j}$, for each cluster is also dependent on the traffic type. Now, the distortion of the data set, $\mathbf{X}$, from each traffic type is given by

$$D(s) = \sum_{i=1}^{N} dist(\mathbf{x_i}, s) \qquad 1 \le s \le S . \tag{4}$$

where S is the number of traffic types. The traffic type, T, of the given data set is identified as the one to which the distortion is minimum,

$$T = \arg\min_s D(s) . \tag{5}$$

5   Bayesian Classification Using Gaussian Mixtures

Traffic classes can also be modeled using GMM. If each dimension of a d-dimensional vector $\mathbf{x}$ is normally distributed random variable with its own mean and variance, and are independent, their joint density has the form

$$p(\mathbf{x}) = \prod_{i=1}^{d} p(x_i) = \prod_{i=1}^{d} \frac{1}{\sqrt{2\pi}\sigma_i} e^{-\frac{1}{2}((x_i - \mu_i)/\sigma_i)^2} . \tag{6}$$

The multivariate Gaussian distribution function [19] for a d-dimensional vector $\mathbf{x}$ is written as

$$p(\mathbf{x}) = \frac{1}{(2\pi)^{d/2}|\boldsymbol{\Sigma}|^{1/2}} e^{-\frac{1}{2}(\mathbf{x}-\boldsymbol{\mu})^t \boldsymbol{\Sigma}^{-1}(\mathbf{x}-\boldsymbol{\mu})} . \tag{7}$$

where $\boldsymbol{\mu}$ is the d-component *mean vector* and $\boldsymbol{\Sigma}$ is the d-by-d *covariance* matrix[2]. As the d components of the vector are independent, the covariance matrix reduces to a diagonal matrix,

---

[2] Superscript t denotes the transpose, and $|\boldsymbol{\Sigma}|$ and $\boldsymbol{\Sigma}^{-1}$ are the determinant and the inverse of the covariance matrix respectively.

$$\mathbf{\Sigma} = \begin{bmatrix} \sigma_1^2 & 0 & \dots & 0 \\ 0 & \sigma_2^2 & \dots & 0 \\ \cdot & \cdot & \cdot & \cdot \\ \cdot & \cdot & \cdot & \cdot \\ \cdot & \cdot & & \cdot \\ 0 & 0 & \dots & \sigma_d^2 \end{bmatrix} . \tag{8}$$

$\sigma_1, \sigma_2, .., \sigma_d$ being the standard deviation along each of the vector component. In our work, the components of the vectors are nothing but the packet train parameters which are statistically independent.

The number of mixtures in a traffic type is a function of the traffic class s, denoted as $n(s)$. Let $\theta_1, \theta_2, ..., \theta_{n(s)}$ be the vectors corresponding to the mixtures $m_1, m_2, ..., m_{n(s)}$, where θ_i is the vector with components μ_i and $\mathbf{\Sigma}_i$ of the mixture m_i. Given the feature vector, $\mathbf{x}$, the Bayes formula [19] to determine the (posteriori) probability of $\mathbf{x}$ being in the ith mixture, m_i, is given as

$$P(\theta_i|\mathbf{x}) \approx P(m_i)p(\mathbf{x}|\theta_i) . \tag{9}$$

An estimate of the probability of a mixture (prior) is calculated as

$$P(m_i) = \frac{n_{m_i}}{N} . \tag{10}$$

where n_{m_i} is the number of vectors in m_i, and N is the total number of vectors.

The Bayes formula can be re-written by substituting the Gaussian function for the *likelihood* of m_i with respect to $\mathbf{x}$

$$P(\theta_i|\mathbf{x}) \approx P(m_i)\frac{1}{(2\pi)^{d/2}|\mathbf{\Sigma}_i|^{1/2}}e^{-\frac{1}{2}(\mathbf{x}-\mu_i)^t\mathbf{\Sigma}_i^{-1}(\mathbf{x}-\mu_i)} . \tag{11}$$

where μ_i is the mean vector and $\mathbf{\Sigma}_i$ is the covariance matrix of mixture m_i.

5.1 Training

We use (9) to determine the probability for each vector in the data set. The parameters (dimensions of the vector) are *packet train length* and *packet train size*. The following algorithm is used in the training phase.

1. Initialize the prior probability of each mixture with equal value such that their sum is 1.
2. Initialize the mean (vector) of each mixture by selecting a vector randomly, such that no two mixtures have the same mean.
3. Initialize the covariance matrix of each mixture to the d-by-d identity matrix.
4. Until the mean and variance of mixtures converge
   – For each vector in the given data set, classify it into mixture m_i if

$$P(\theta_i|\mathbf{x}) > P(\theta_j|\mathbf{x}) \quad \text{for all } j \neq i . \tag{12}$$

   where θ_i corresponds to mixture m_i. If there are two or more mixtures with maximum probabilities, one among them is chosen arbitrarily.

- Recompute the probability of each mixture.
- Recompute the mean vector of each mixture.
- Recompute the covariance matrix of each mixture.

The above algorithm basically computes the vector θ_i and the probability corresponding to every mixture of a traffic class.

5.2   Testing

The testing phase uses the probabilities and the means and variances of mixtures of different traffic classes obtained from the training phase. In Bayesian clustering, the probability that the ith vector, $\mathbf{x_i}$ belongs to a traffic class s is found using

$$P(s|\mathbf{x_i}) = \sum_j P(\theta_j|\mathbf{x_i}) \qquad 1 \leq j \leq n(s) \ . \tag{13}$$

where $n(s)$ is the number of mixtures in the traffic class denoted by s.

We have observed that the rule which includes only the maximum probability corresponding to the best mixture, yields better results as compared to the above rule. We have therefore chosen the probability of occurrence of a vector in a traffic class to be the maximum of the probabilities of occurrence of the vector in each of the mixtures in the particular traffic class. That is

$$P(s|\mathbf{x_i}) = \max_j P(\theta_j|\mathbf{x_i}) \qquad 1 \leq j \leq n(s) \ . \tag{14}$$

The probability that the given set of vectors belong to a particular traffic class is determined by the joint probability of all the vectors

$$P(s) = \prod_{i=1}^{N} P(s|\mathbf{x_i}) \ . \tag{15}$$

where N is the total number of vectors.

Traffic classes overlap and the posterior probabilities obtained for mixtures across different traffic classes can be similar. We therefore apply a threshold on the probabilities (obtained by experimentation) before assigning a particular class. The thresholds are different for each of the traffic classes. For example, POP3 and SMTP have similar characteristics. A threshold based classification reduces the misclassification of POP3 as SMTP. The traffic class of the given data set is based on the following decision:

$$T = \arg\max_s \left[P(s) : P(s) > \alpha(s)\right] \ . \tag{16}$$

where $\alpha(s)$ is the threshold for traffic class, s.

6   Evaluation

The packet traces were collected from a gateway connecting the TeNeT [20] private LAN and Internet Service Provider's 1 Mbps (megabits per second) link, for a period of 90 days using tcpdump [13]. The packet header information from the tcpdump output is fed to a parser program. The parser generates data sets by computing the packet train parameters of each traffic type based on packet headers information (such as source IP address, destination IP address, source port, destination port, packet length and TCP flags). A data set used during training consists of packet train parameters of a single traffic type. The training was performed on data taken from 60 days of real Internet traffic. Data sets for varying time periods of 60 minutes, 30 minutes and 15 minutes were used for training. This essentially means that one-hour model for the ith hour has packet train information pertaining to that hour of the 60 days data. Such models are generated for each of the traffic types. So an HTTP data set for 12th hour has 60 days of HTTP data for the same hour. The models generated in the training phase were used for classification of the given data set during testing. Testing was performed on Internet data collected over 30 days. Data sets of 60 minutes, 30 minutes and 15 minutes were used for testing. The data collection interval used for obtaining training data is same as for obtaining test data, so that the data sets are subject to the same biases.

Classification using VQ was tested with one hour models. Bayesian classification using Gaussian mixtures was tested using models of one hour, 30 minutes and 15 minutes; and the results were compiled separately. It should be noted that the actual test data used are obtained by extracting packet trains for a connection. The testing program will predict the traffic type for the given set of packet trains. The correctness of classification is determined by comparing the predicted traffic type with the actual traffic type. Accuracy is the percentage of data sets predicted correctly using the models.

It was observed that the number of mixtures (or clusters) required to model a particular traffic class depends on the application protocol. This number is manually found out by comparing the mean and variance of a mixture with the means and variances of other mixtures. For this, the training algorithms used in both the approaches (VQ and GMM) were executed for different number of clusters/mixtures to find the optimal number of clusters/mixtures required. The number of mixtures required to model different traffic classes were found to be different. For example, HTTP traffic was modeled using 11 different Gaussian mixtures, whereas POP3 traffic required only 7 Gaussian mixtures to model it.

6.1   Results Using VQ

Table 1 shows the accuracy in classifying traffic using VQ. The results of classification with and without thresholds for traffic classes are shown in Table 1(a) and Table 1(b) respectively. By defining thresholds for the distortion of each traffic type, misclassification error has reduced. The overall performance of classification increased from 83.3% accuracy to 94.9% accuracy.

6.2   Results Using GMM

Table 2 shows the result of classifying traffic using the GMM. An improvement in classifying traffic (from 96.9% accuracy to 98.6% accuracy) was observed when

Table 1. Results using VQ for one hour data

(a) Without any threshold

Traffic Type	Accuracy
HTTP	98.55%
SMTP	81.16%
DNS	100%
POP3	76.08%
SSH	60.86%

(b) With thresholds defined

Traffic Type	Accuracy
HTTP	99.27%
SMTP	96.38%
DNS	100%
POP3	90.56%
SSH	88.40%

Table 2. Results using GMM for one hour data

(a) Without any threshold

Traffic Type	Accuracy
HTTP	99.60%
SMTP	99.30%
DNS	100%
POP3	95.90%
SSH	79.18%

(b) With thresholds defined

Traffic Type	Accuracy
HTTP	99.60%
SMTP	99.30%
DNS	100%
POP3	97.20%
SSH	96.92%

the thresholds are defined. Also, classification using multiple parameters, namely packet train length and packet train size, has given more accurate results when compared to the previous work using Bayesian analysis techniques [8] which obtained an accuracy of 95% using 28 minutes data. Comparing tables 1 and 2 shows that modeling using GMM gives more accurate classification results than modeling using VQ. Test results in Table 2 was obtained with traffic for one hour slots. Tables 3 and 4 show results with thresholds defined, using 30 minutes data and 15 minutes data respectively.

6.3 Discussion

It is evident from the test results that the DNS traffic is always identified correctly. DNS is an Internet service which uses UDP as the underlying transport protocol. Since UDP is a connectionless protocol, each DNS query or reply will be a packet train with single packet. This clearly distinguishes it from TCP flows which will normally have more than one packet in a train. Due to this unique packet train property of the DNS traffic, it is never misclassified.

Tables 2, 3 and 4 show that HTTP and SMTP traffic types were classified with almost the same accuracies for different time durations even without thresholds. The mixtures of HTTP traffic type did not overlap with mixtures of any other traffic types resulting in highly accurate classification. SMTP and POP3 traffic had two overlapping mixtures; therefore POP3 got misclassified as SMTP traffic. This is expected as both POP3 and SMTP deal with mail traffic. The characteristics of both types of traffic are expected to be similar.

One important feature of these models generated using packet train parameters is that it can be used by network administrators to estimate the number of

Table 3. Results using GMM for 30 minutes data with thresholds defined

Traffic Type	Accuracy
HTTP	99.84%
SMTP	99.68%
DNS	100%
POP3	96.99%
SSH	96.72%

Table 4. Results using GMM for 15 minutes data with thresholds defined

Traffic Type	Accuracy
HTTP	99.78%
SMTP	99.67%
DNS	100%
POP3	96.28%
SSH	94.53%

Table 5. Relevant HTTP mixtures during 11:00-12:00

Index	Packet train length		Packet train size		Percentage of total trains
#	μ	σ	μ	σ	%
1.	5.10	1.02	854.52	188.30	20.24
2.	10.54	3.0	7428.57	4485.69	18.85
3.	4.81	0.41	576.14	27.58	18.27
4.	4.67	0.67	436.87	39.34	13.00
5.	34.56	18.02	37356.1	24876.7	11.90
6.	6.06	0.74	2644.25	545.22	6.93
7.	5.21	1.11	1522.81	163.07	5.51
8.	3.56	0.97	243.79	92.40	2.70
9.	520.56	1582.78	619126.8	2105716.6	1.18

tiny flows, that is ones that involve only a small number of packets. Similarly, administrators can also monitor *heavy hitters* [21], those that represent a significant proportion of the traffic, or the link bandwidth. For example, consider some of the relevant HTTP mixtures for inbound packet trains during 11-12 hour from the TeNeT link, which are shown in table 5. The table shows that most of the HTTP downloads are small packet trains. As seen in rows 1, 3, and 4 of the table, more than 50% of downloads are trains consisting of 5 packets, with the download size ranging approximately from 400 to 1000 bytes. Near to 12% of download traffic are medium-size downloads as seen in row 5, and more than 1% are huge downloads as depicted in the last row of the table. Such information can be used to learn the different classes of service required for various applications in the network. The knowledge of traffic characteristics enables an administrator to do QoS provisioning and also to prioritize different traffic classes depending on the QoS requirements.

This statistical method also reveals typical use of an application. For example, if HTTP were used for relaying some other traffic (such as SSH) or for streaming large amount media traffic, rather than more traditional web browsing. Also traffic traces were collected during file transfer between machines using SCP, which uses the same port as SSH (port 22). The SCP packet trains generated by transferring files of size greater than 2 Megabytes was successfully rejected by the SSH model. Though SCP and SSH use the same port, SSH traffic consists of small packets (of size between 100 bytes to 200 bytes), whereas SCP traffic usually has large packets (of size usually greater than 1000 bytes).

7   Conclusion

Our experiments show that GMM can be used to generate better models as compared to VQ. The accuracy of Bayesian classification for one hour data using GMM is 98.6% compared to 94.9% achieved using VQ based classification. Accuracy of classification using GMM for data of 15 minutes is 98.05%. The significance of time duration is that, only 15 minutes of data is required to identify a particular traffic type. Although VQ and GMM have been used for modeling of network traffic in the past [8,10], a novelty of the work presented in this paper is the use of new parameters, namely *packet train length* and *packet train size*. Using *packet train length* and *packet train size* as multiple parameters for modeling, has helped not only in yielding high accuracy in classification, but also in revealing useful information on different service classes in network traffic.

This work looked only at one UDP based traffic class, namely DNS traffic. But, it should be noted that almost all UDP based applications can be viewed as a connection oriented traffic and therefore can be represented using packet trains. For example, a video conferencing tool running on top of UDP can be identified uniquely by *src host*, *src port*, *dst host* and *dst port*, and all such packets can be considered as part of a connection. To distinguish different connections of the same applications, the time between packets can be used; as the time between consecutive packets of one run of the application will be much less as compared to time between consecutive packets of different runs of the application. Hence, using all these information, packet train parameters can be extracted for the traffic generated by a video conferencing application, or in general, for traffic generated by almost any UDP based application.

References

1. Croll, A., Packman, E.: Managing Bandwidth: Deploying Across Enterprise Networks. Prentice Hall PTR Internet Infrastructure Series. Prentice Hall (2000)
2. Sally Floyd and Vern Paxson: Difficulties in simulating the Internet. IEEE/ACM Trans. on Networking **9** (2001) 392–403
3. Logg, C.: Characterization of the traffic between SLAC and the Internet. *http://www.slac.stanford.edu/comp/net/slac-netflow/html/SLAC-netflow.html* (2003)

4. Moore, A.W., Papagiannaki, K.: Toward the accurate identification of network applications. In: Passive and Active Network Measurement, Sixth International Workshop, PAM 2005. (2005) 41–54
5. Roughan, M., Sen, S., Spatscheck, O., Duffield, N.G.: Class-of-service mapping for QoS: a statistical signature-based approach to IP traffic classification. In: Internet Measurement Conference, IMC '04. (2004) 135–148
6. McGregor, A., Hall, M., Lorier, P., Brunskill, J.: Flow Clustering Using Machine Learning Techniques. In: Passive and Active Network Measurement, Fifth International Workshop, PAM 2004. (2004) 205–214
7. Chen, Y.W.: Traffic behavior analysis and modeling of sub-networks. International Journal of Network Management 12 (2002) 323–330
8. Moore, A.W., Zuev, D.: Internet traffic classification using bayesian analysis techniques. In: Proc. of the 2005 ACM SIGMETRICS, International Conference on Measurement and Modeling of Computer Systems, ACM Press (2005) 50–60
9. Saifulla, M.A., Murthy, H.A., Gonsalves, T.A.: Identifying Patterns in Internet Traffic. In: International Conference on Computer Communication. (2002) 859–865
10. M. A. Saifulla: A Pattern Matching Approach To Classification Of Internet Traffic. Master's thesis, Indian Institute of Technology, Madras (2003)
11. Tamir, D., yeon Park, C., Yoo, W.S.: Vector Quantization and Clustering: A Pyramid Approach. IEEE Data Compression Conference and Industrial Workshop (1995)
12. Roberts, S.J., Husmeier, D., Penny, W., Rezek, I.: Bayesian Approaches to Gaussian Mixture Modelling. IEEE Trans. on Pattern Analysis and Machine Intelligence 20 (1998) 1133–1142
13. Stevens, W.R.: TCP/IP Illustrated, Volume 1: The Protocols. Addison-Wesley (1994)
14. IANA: Internet Assigned Numbers Authority. http://www.iana.org/assignments/port-numbers (2006)
15. Claffy, K.C.: Internet traffic classification. PhD thesis, University of California, San Diego (1994)
16. Paxson, V.: Empirically derived analytic models of wide-area TCP connections. IEEE/ACM Trans. on Networking 2 (1994) 316–336
17. Jain, R., Routhier, S.: Packet Trains-Measurements and a New Model for Computer Network Traffic. IEEE Journal on Selected Areas in Communications SAC-4 (1986) 986–995
18. M.Gray, R.: Vector Quantization. IEEE ASSP Magazine 1 (1984) 4–29
19. Duda, R.O., Hart, P.E., Stork, D.G.: Pattern Classification. Second edn. Wiley-Interscience Publication (2001) Chapter 2.
20. TeNeT: The Telecommunications and Computer Networking Group, Indian Institute of Technology, Madras. http://www.TeNeT.res.in (1996)
21. Graham Cormode and Flip Korn and S. Muthukrishnan and Divesh Srivastava: Diamond in the Rough: Finding Hierarchical Heavy Hitters in Multi-Dimensional Data. In: Proc. of the ACM SIGMOD, International Conference on Management of Data. (2004) 155–166

Adaptive Bandwidth Allocation Method for Long Range Dependence Traffic

Bong Joo Kim and Gang Uk Hwang*

Division of Applied Mathematics and Telecommunication Program
Korea Advanced Institute of Science and Technology
373-1 Guseong-dong, Yuseong-gu, Taejeon, 305-701, South Korea
KimBongJoo@kaist.ac.kr
guhwang@amath.kaist.ac.kr
http://queue.kaist.ac.kr/~guhwang

Abstract. In this paper, we propose a new method to allocate bandwidth adaptively according to the amount of input traffic volume for a long range dependent traffic requiring Quality of Service (QoS). In the proposed method, we divide the input process, which is modelled by an $M/G/\infty$ input process, into two sub-processes, called a long time scale process and a short time scale process. For the long time scale process we estimate the required bandwidth using the linear prediction. Since the long time scale process varies (relatively) slowly, the required bandwidth doesn't need to be estimated frequently. On the other hand, for the short time scale process, we use the large deviation theory to estimate the effective bandwidth of the short time scale process based on the required QoS of the input traffic. By doing this we can capture the short time scale fluctuation by a buffer and the long time scale fluctuation by increasing or decreasing the bandwidth adaptively. Through simulations we verify that our proposed method performs well to satisfy the required QoS.

1 Introduction

Several traffic measurement studies in recent years have shown the existence of long-range dependence(LRD) and self-similarity in network traffic such as Ethernet LANs[6,11], variable bit rate(VBR) video traffic[2,7], Web traffic[5], and WAN traffic[15]. In addition, many analytical studies have shown that LRD network traffic can have a detrimental impact on network performance and pointed out the need of revisiting various issues of performance analysis and network design. One of the practical impacts of LRD is that buffers at switches and multiplexers should be significantly larger than those predicted by traditional queueing analysis and simulations to meet the required quality of service (QoS). This requirement for large buffers can be explained by the Noah effect and the

---

* This work was supported by Korea Research Foundation Grant funded by Korea Government (MOEHRD, Basic Research Promotion Fund) (KRF-2005-003-C00022).

G. Parr, D. Malone, and M. Ó Foghlú (Eds.): IPOM 2006, LNCS 4268, pp. 13–24, 2006.

Joseph effect. However, such a significantly large buffer requirement causes the inefficient use of the network resources.

To solve this problem, many studies recommend that the buffer be kept small while the link bandwidth is to be increased, and most studies, e.g. [3,9,16,17] in the open literature have been focusing on the estimation of the required bandwidth for LRD traffic based on the buffer overflow probability. However, a deterministic bandwidth allocation strategy seems not to be quite effective for LRD traffic. For example, if the traffic volume in a certain (relatively long) period of time is less than the estimated bandwidth, which can happen for LRD traffic with nonnegligible probability, then the network wastes the bandwidth for as much period of time. Therefore, it would be more effective if we can adaptively change the bandwidth allocated for LRD traffic, which is the motivation of this study.

In this paper, we propose a new method to allocate bandwidth adaptively according to the amount of input traffic volume for a long range dependent traffic requiring quality of service (QoS). We consider a discrete time queueing system with an $M/G/\infty$ input process[10,12,13,14] to model the system with LRD traffic. The $M/G/\infty$ input process is the busy server process of a discrete time infinite server system fed by Poisson arrivals of rate λ (customers/slot) with general service times. To mimic LRD traffic the discretized Pareto distribution is considered as the service time distribution.

In the proposed method, We divide the input process into two sub-processes: a long time scale process and a short time scale process. For the long time scale process we estimate the required bandwidth using the minimum mean square error(MMSE) linear prediction. Since the long time scale process varies (relatively) slowly, the required bandwidth doesn't need to be estimated frequently. On the other hand, for the short time scale process, since the required bandwidth for the long time scale process is fully captured by the linear prediction, we use the large deviation theory [4] to estimate the effective bandwidth of the short time scale process based on the required QoS of the input traffic. By doing this we can capture the short time scale fluctuation by a buffer and the long time scale fluctuation by increasing or decreasing the bandwidth adaptively. Hence, the total bandwidth allocated for the $M/G/\infty$ input process is obtained by adding the effective bandwidth of the short time scale process to the estimated bandwidth of the long time scale process.

To check the performance of our proposed method, we simulate a queueing system with the $M/G/\infty$ input process under various conditions and our simulation results show that the proposed method performs well. Further discussion on the proposed method will be given later.

The organization of this paper is as follows: In section 2, we introduce a stationary $M/G/\infty$ input process and investigate how to model LRD traffic with the stationary $M/G/\infty$ input process. In section 3, we propose our adaptive bandwidth allocation method for LRD traffic. In section 4, we present numerical results to validate our method and further discussions are also provided. Finally, we have conclusions in section 5.

2 The Stationary $M/G/\infty$ Input Process

An $M/G/\infty$ input process is the busy server process of a discrete-time infinite server system fed by Poisson arrivals with general service time. This process has been known to be useful to model LRD traffic because it is stable under multiplexing and flexible in capturing positive dependencies over a wide range of time scales[10,12,13,14]. To describe the $M/G/\infty$ input process, we assume that the time axis is divided into slots of equal size and consider a discrete time $M/G/\infty$ queueing system as follows: The customer arrival process is according to a Poisson process with rate λ, and β_{t+1} new customers arrive into the system during time slot $[t, t+1)$ for $t = 0, 1, \cdots$. Since the number of servers is infinite, customer j, $j = 1, ..., \beta_{t+1}$, immediately occupies a new server in the system after its arrival and begins its service from slot $[t+1, t+2)$, with service time $\sigma_{t+1,j}$ (slots). We assume that $\{\sigma_{t,j}, t = 1, 2, \cdots; j = 1, 2, \cdots \beta_t\}$ are i.i.d. and σ denotes a generic r.v. for $\sigma_{t,j}$.

Let $X(t)$, $t \geq 0$ denote the number of busy servers during time slot $[t, t+1)$, in other words, the number of customers still present in the $M/G/\infty$ system during time slot $[t, t+1)$. We assume that the system starts with $X(0)$ initial customers at time 0. Then, the $M/G/\infty$ input process is the busy server process $\{X(t), t = 0, 1, \cdots\}$ of the $M/G/\infty$ system described above. When $X(t)$ is used for traffic modelling, $X(t)$ is considered as the number of packets arriving during slot $[t, t+1)$.

When we assume that the initial number $X(0)$ of customers is a Poisson r.v. with parameter $\lambda E[\sigma]$, and that $\{\sigma_{0,j}, j = 1, 2, \cdots, X(0)\}$ are i.i.d. $\mathbb{N}$-valued r.v.s distributed according to the forward recurrence time $\hat{\sigma}$ associated with σ, i.e., the p.m.f. of $\hat{\sigma}$ is given by $p[\hat{\sigma} = r] \triangleq \frac{P[\sigma \geq r]}{E[\sigma]}, r = 1, 2, \cdots$, it can be shown that the resulting $M/G/\infty$ input process is stationary[10,13]. In addition, we can show that the covariance structure of $\{X(t), n = 0, 1, \cdots\}$ is given by [13]

$$\Gamma(k) \triangleq cov[X(t), X(t+k)] = \lambda E\left[(\sigma - k)^+\right], \quad t, k = 0, 1, \cdots, \tag{1}$$

which is further reduced as [10,13,18]

$$\Gamma(k) = \lambda \sum_{i=0}^{\infty} P\left[(\sigma - k)^+ > i\right] = \lambda \sum_{i=k+1}^{\infty} P[\sigma \geq i] = \lambda E[\sigma]P[\hat{\sigma} > k]. \tag{2}$$

In addition, the ACF (autocorrelation function) $\rho(k)$ of the stationary $M/G/\infty$ input process is obtained from (1) and (2) as follows:

$$\rho(k) \triangleq \frac{\Gamma(k)}{\Gamma(0)} = P[\hat{\sigma} > k], \quad k = 0, 1, \cdots.$$

Next, we use the stationary $M/G/\infty$ input process to model LRD traffic. To do this, we consider a discretized Pareto distribution as the service time σ of a customer in the corresponding $M/G/\infty$ queueing system. The discretized Pareto distribution is defined by

$$P[\sigma = i] \triangleq P[i \leq Y < i+1] \tag{3}$$

where the random variable Y has the Pareto distribution with the shape parameter $1 < \gamma < 2$ and the location parameter $\delta(> 0)$, given by

$$P[Y \leq x] = \begin{cases} 1 - (\frac{\delta}{x})^\gamma & \text{if } x > \delta, \\ 0 & \text{otherwise.} \end{cases}$$

Then, it can be easily checked that the stationary $M/G/\infty$ input process with discretized Pareto service times given above exhibits a long range dependence by using (1),(2) and (3). We omit the detailed proof in this paper due to the limitation of space. Hence, we use the the stationary $M/G/\infty$ input process with discretized Pareto service times to mathematically model LRD traffic.

3   Adaptive Bandwidth Allocation Method

In this section, we propose a new adaptive bandwidth allocation (ABA) method for LRD traffic requiring quality of service (QoS). In our ABA method, we assume that the LRD traffic is modelled by a stationary $M/G/\infty$ input process with discretized Pareto service times. To compute the bandwidth allocated for LRD traffic, we decompose the $M/G/\infty$ input traffic $X(t)$ into two components called a short time scale process, denoted by $X_s(t)$, and a long time scale process, denoted by $X_l(t)$ as follows: To decompose the input LRD traffic, a threshold value T is given *a priori*. Then, $X_s(t)$ is the number of busy servers at slot $[t-1, t]$ which became active by arrivals whose service times are less than or equal to a given threshold T, and $X_l(t)$ is the number of busy servers at slot $[t-1, t]$ which became active by arrivals whose service time is greater than T.

Due to the independence decomposition property of a Poisson process with respect to a random selection, we see that the Poisson arrivals with service times less than or equal to T generate $X_s(t)$ and accordingly, $X_s(t)$ is an $M/G/\infty$ input process with arrival rate $\lambda P[\sigma \leq T]$ and service time with p.m.f $\frac{P[\sigma=k]}{P[\sigma \leq T]}, 1 \leq k \leq T$. Similarly, $X_l(t)$ is also an $M/G/\infty$ input process with arrival rate $\lambda P[\sigma \geq T+1]$ and service time with p.m.f $\frac{P[\sigma=k]}{P[\sigma \geq T+1]}, k \geq T+1$. In addition, we see that both processes $X_l(t)$ and $X_s(t)$ are independent. Then, by adjusting the initial numbers and service times of customers for both processes, we make both processes stationary. From the definitions of $X_s(t)$ and $X_l(t)$, we see that $X_s(t)$ captures the short time fluctuation in the input LRD traffic and $X_l(t)$ captures the long time fluctuation in the input LRD traffic.

Our next step is to compute the effective bandwidths for $X_s(t)$ and $X_l(t)$ based on the given QoS requirement of the input LRD traffic. Since the service times of customers generating the short time scale process $X_s(t)$ are bounded by T, we can easily show that $X_s(t)$ is a short range dependence process. Hence, we use the large deviation theory [4] to compute the effective bandwidth of $X_s(t)$ based on the QoS requirement of the input LRD traffic. We will give the details in subsection 3.1. For the long time scale process $X_l(t)$, since the tail behavior of the service times of customers generating $X_l(t)$ is the same as the discretized Pareto service times, we see that $X_l(t)$ is a LRD traffic. However, since the service times of customers generating $X_l(t)$ are relatively long, the process $X_l(t)$ is changing relatively slowly, so that we may think that $X_l(t)$ is almost constant during each time period of fixed length, called the basic allocation period. In addition, since the required QoS is already considered in the computation of the effective bandwidth of $X_s(t)$, we predict the amount of $X_l(t)$ based on the previous history of $X_l(t)$ at the first slot of each basic allocation period and allocate the same amount of bandwidth as the prediction for $X_l(t)$ during the basic allocation period. We will use the minimum mean square error (MMSE) linear predictor to predict the amount of $X_l(t)$, which will be given in detail in subsection 3.2. Finally, we will describe how to allocate the bandwidth for the LRD input traffic adaptively based on the effective bandwidth of $X_s(t)$ and the estimated bandwidth of $X_l(t)$ in subsection 3.3.

3.1   The Effective Bandwidth for $X_s(t)$

In this subsection, we estimate the effective bandwidth for the short time scale process $X_s(t)$. To do this, we assume that the system has a buffer of size b and should guarantee the overflow probability less than $e^{-\xi b}$ for the input LRD process. As mentioned above, since the effective bandwidth function of $X_l(t)$ is predicted by a MMSE linear predictor which can not consider the QoS requirement, we consider the QoS requirement of the input LRD process in the computation of the effective bandwidth of $X_s(t)$, which is given in the following theorem:

Theorem 1. *When b is the buffer size of the system, the effective bandwidth C_s for the required overflow probability $e^{-\xi b}$ is given by*

$$C_s = \sum_{n=1}^{T} \lambda P[\sigma = n] \frac{e^{\xi n} - 1}{\xi}.$$

PROOF: First, note that the arrival process of customers generating $X_s(t)$ can be decomposed into T independent Poisson processes with rate $\lambda_n = \lambda P[\sigma = n]$. For convenience, the Poisson process with rate λ_n is called the λ_n-Poisson process. Let $B_{n,i}$ be the number of customers arriving in $[i-1, i)$ and $A_n(t)$ be the total amount of traffic arrived in $[0, t)$ for the λ_n-Poisson process. Then, observing that each customer of the λ_n-Poisson process eventually generates n packets (since the service time of the customer is always n), for sufficiently large t we get

$$A_n^{(l)}(t) \triangleq \tilde{B}_0 + \sum_{i=1}^{t-n} B_{n,i} \cdot n \le A_n(t) \le \tilde{B}_0 + \sum_{i=1}^{t} B_{n,i} \cdot n \triangleq A_n^{(u)}(t), \quad (4)$$

where $\tilde{B}_0$ denotes the amount of traffic due to the initial customers of the λ_n-Poisson process. Note that $B_{n,i}$ are i.i.d poisson random variables with parameter λ_n. From the first inequality of (4) we get

$$E[e^{\theta A_n(t)}] \ge E[e^{\theta \sum_{i=1}^{t-n} B_{n,i} \cdot n}] = E[\Pi_{i=1}^{t-n} e^{\theta B_{n,i} \cdot n}] = \Pi_{i=1}^{t-n} E[e^{\theta B_{n,i} \cdot n}].$$

By taking the logarithm we obtain,

$$\log E[e^{\theta A_n(t)}] \geq \sum_{i=1}^{t-n} \log E[e^{\theta \cdot n B_{n,i}}] = (t-n)\lambda_n(e^{\theta \cdot n} - 1)$$

since B_n is the poisson r.v. with parameter λ_n. Then it follows that

$$\lim_{t \to \infty} \frac{1}{t} \log E[e^{\theta A_n(t)}] \geq \lambda_n(e^{\theta \cdot n} - 1).$$

Similarly, from the second inequality of (4) we can show that $\lim_{t \to \infty} \frac{1}{t} \log E[e^{\theta A_n(t)}] \leq \lambda_n(e^{\theta \cdot n} - 1)$. Hence, the Gärtner-Ellis limit [4] of the λ_n-Poisson process is given by

$$\lim_{t \to \infty} \frac{1}{t} \log E[e^{\theta A_n(t)}] = \lambda_n(e^{\theta \cdot n} - 1).$$

Now, since $X_s(t)$ is the superposition of the λ_n-Poisson processes for $1 \leq n \leq T$, the Gärtner-Ellis limit of $X_s(t)$ is given by $\Lambda(\theta) = \sum_{n=1}^{T} \lambda_n(e^{\theta \cdot n} - 1)$.

Therefore, when b is buffer size, the effective bandwidth C_s for $X_s(t)$ with the overflow probability $e^{-\xi b}$ can be calculated as follows [4]:

$$C_s = \frac{\Lambda(\xi)}{\xi} = \sum_{n=1}^{T} \lambda P[\sigma = n] \frac{e^{\xi n} - 1}{\xi},$$

■

3.2   The Effective Bandwidth for $X_l(t)$

In this subsection, we estimate the effective bandwidth for the long time scale process $X_l(t)$. Even though we can't extract $X_l(t)$ from $X(t)$ directly, we can achieve our purpose by introducing a new process $Z(t)$ defined by

$$Z(t) \triangleq X(t) - E[X_s(t)] = X_l(t) + X_s(t) - E[X_s(t)]$$
$$= X_l(t) + \eta(t),$$

where $\eta(t) = X_s(t) - E[X_s(t)]$. The process $\eta(t)$ is viewed as a noise process which is not i.i.d., but $E[\eta(t)] = 0$. Note that $X_l(t)$ and $\eta(t)$ are independent because $X_l(t)$ and $X_s(t)$ are independent. In addition, if we know the threshold value T and the input traffic parameters λ, γ and δ, we can get $Z(t)$ from $X(t)$.

Next, we consider the pth order linear predictor which can estimate $\hat{X}_l(t+1)$ of $X_l(t+1)$ using a linear combination of the current and previous values of $X_l(t)$ as follows:

$$\hat{X}_l(t+1) \triangleq \sum_{i=0}^{p-1} \omega(i) Z(t-i), \quad \sum_i \omega(i) = 1, \tag{5}$$

where the coefficients $\omega(i)$ can be obtained by an induced linear system given below. The optimal linear predictor in the mean square sense is such that minimizes the mean square error $\psi = E\left[\left(X_l(t+1) - \hat{X}_l(t+1)\right)^2\right]$. Note that ψ is represented by a function of the vector $\boldsymbol{\omega} = (\omega(0), \omega(1), \cdots, \omega(p-1))$. So, the vector $\boldsymbol{\omega}$ that minimizes ψ is found by taking the gradient, setting it equal to zero and then solving it for $\boldsymbol{\omega}$. This results in

$$E\left[X_l(t+1)Z(t-j)\right] = E\left[\hat{X}_l(t+1)Z(t-j)\right]. \tag{6}$$

From the facts that $X_l(t)$ and $\eta(t)$ are independent and $E\left[\eta(t)\right] = 0$ for all t, the left hand side of (6) is computed as follows:

$$\begin{aligned} E\left[X_l(t+1)Z(t-j)\right] &= E\left[X_l(t+1)(X_l(t-j)+\eta(t-j))\right] \\ &= r(j+1), \quad j = 0, 1, \cdots, p-1, \end{aligned} \tag{7}$$

where $r(k) = E\left[X_l(t)X_l(t+k)\right]$. Similarly, the right hand side of (6) is computed as follows:

$$\begin{aligned} E\left[\hat{X}_l(t+1)Z(t-j)\right] &= E\left[\sum_{i=0}^{p-1} \omega(i)Z(t-i)Z(t-j)\right] \\ &\quad -\sum_{i=0}^{p-1} \omega(i)E\left[X_l(t-i)X_l(t-j)\right] + \sum_{i=0}^{p-1} \omega(i)E\left[\eta(t-i)\eta(t-j)\right] \\ &= \sum_{i=0}^{p-1} \omega(i)\left[r(j-i)+s(j-i)\right], \quad 0 \le j \le p-1, \end{aligned} \tag{8}$$

where $s(k) = E\left[\eta(t)\eta(t+k)\right]$. Combining (6),(7) and (8) yields

$$r(j+1) = \sum_{i=0}^{p-1} \omega(i)\left[r(j-i)+s(j-i)\right], \quad j = 0, 1, \cdots, p-1.$$

In matrix form, this leads to the following linear system, so called Weiner-Hopf linear equation [1,8] whose solution gives the optimum filter coefficients $\omega(0), \cdots, \omega(p-1)$:

$$\begin{bmatrix} r(1) \\ r(2) \\ r(3) \\ \vdots \\ r(p) \end{bmatrix} = \begin{bmatrix} r(0)+s(0) & r(1)+s(1) & \cdots & r(p-1)+s(p-1) \\ r(1)+s(1) & r(0)+s(0) & \cdots & r(p-2)+s(p-2) \\ r(2)+s(2) & r(1)+s(1) & \cdots & r(p-3)+s(p-3) \\ \vdots & \vdots & \ddots & \cdots \\ r(p-1)+s(p-1) & r(p-2)+s(p-2) & \cdots & r(0)+s(0) \end{bmatrix} \begin{bmatrix} \omega(0) \\ \omega(1) \\ \omega(2) \\ \vdots \\ \omega(p-1) \end{bmatrix}. \tag{9}$$

It remains to compute $r(k)$ and $r(k)+s(k)$ to get $\omega(i), 0 \le i \le p-1$, which result in the optimal linear predictor. We need the following theorem:

Theorem 2. $r(k)$ and $r(k)+s(k)$ are given by

$$r(k) = \Gamma^*(k) + E^2\left[X_l(t)\right],$$
$$r(k)+s(k) = \Gamma(k) + E^2\left[X_l(t)\right],$$

where $\Gamma^*(k)$ is autocovariance of lag k for $X_l(t)$ and given by

$$\Gamma^*(k) = \begin{cases} \lambda \left[E\left[\sigma\right] - k + \sum_{t=1}^{k}(k-t)P\left[\sigma = t\right] \right], & \text{if } k \geq T, \\ \lambda \left[P\left[\sigma \geq T+1\right](T-k) + E\left[\sigma\right] - T + \sum_{t=1}^{T}(T-t)P\left[\sigma = t\right] \right], & \text{if } k \leq T-1. \end{cases} \tag{10}$$

and $\Gamma(k)$ is autocovariance of lag k for $X(t)$, given in (2).

PROOF: To compute $r(k)$ we only need to compute $\Gamma^*(k)$. Since a long time scale process $X_l(t)$ is an $M/G/\infty$ input process with arrival rate $\lambda P[\sigma \geq T+1]$ and service time σ^* with p.m.f $\frac{P[\sigma=k]}{P[\sigma \geq T+1]}$, $k \geq T+1$, with the help of (2) $\Gamma^*(k)$ can be expressed as follows:

$$\Gamma^*(k) = \lambda P[\sigma \geq T+1]E[\sigma^*]P[\hat{\sigma}^* > k], \tag{11}$$

where $E[\sigma^*]$ is the mean of service times in a long time scale process $X_l(t)$ and $\hat{\sigma}^*$ is the forward recurrence time associated with σ^*. Observe that $P[\hat{\sigma}^* > k]$ can be calculated as follows:

$$P[\hat{\sigma}^* > k] = \sum_{i=k+1}^{\infty} P[\hat{\sigma}^* = i] = \sum_{i=k+1}^{\infty} \frac{P[\sigma^* \geq i]}{E[\sigma^*]}$$

$$= \begin{cases} \frac{1}{E[\sigma^*]} \frac{1}{P[\sigma \geq T+1]} \left[E\left[\sigma\right] - k + \sum_{t=1}^{k}(k-t)P\left[\sigma = t\right] \right], & \text{if } k \geq T, \\ \frac{1}{E[\sigma^*]} \left\{ (T-k) + \frac{1}{P[\sigma \geq T+1]} \left[E\left[\sigma\right] - T + \sum_{t=1}^{T}(T-t)P\left[\sigma = t\right] \right] \right\}, & \text{if } k \leq T-1. \end{cases} \tag{12}$$

Then combining (11) and (12) we show that $\Gamma^*(k)$ is given by (10).

Next, $r(k) + s(k)$ can be computed as follows:

$$\begin{aligned} r(k) + s(k) &= E\left[X_l(t)X_l(t+k)\right] + E\left[\eta(t)\eta(t+k)\right] \\ &= E\left[\{X_l(t) + \eta(t)\}\{X_l(t+k) + \eta(t+k)\}\right] \\ &\quad \text{since } E[\eta(t)] = 0 \text{ for all } t, \text{ and } \eta(t) \text{ and } X_l(t) \text{ are independent} \\ &= E\left[\{X(t) - E\left[X_s(t)\right]\}\{X(t+k) - E\left[X_s(t+k)\right]\}\right] \\ &\quad \text{by the definition of } \eta(t) \\ &= E\left[X(t)X(t+k)\right] - 2E[X(t)]E[X_s(t)] + E^2\left[X_s(t)\right] \\ &\quad \text{by the staionarity of } X(t) \text{ and } X_s(t) \\ &= \Gamma(k) + E^2\left[X(t)\right] - 2E[X(t)]E[X_s(t)] + E^2\left[X_s(t)\right] \\ &= \Gamma(k) + (E[X(t)] - E[X_s(t)])^2 \\ &= \Gamma(k) + E^2[X_l(t)], \end{aligned}$$

where $\Gamma(k)$ is autocovariance of lag k for $X(t)$, given in (2). ∎

Consequently if we have the parameters λ, γ and δ of the $M/G/\infty$ input traffic, which can be obtained from the previously measured data or experience, the coefficients $\{\omega(i)\}_{i=0}^{p-1}$ can be determined with the help of Theorem 2 and (9). Then, by substituting $\omega(i)$ values into equation (5), we can obtain $\hat{X}_l(t)$ as an estimate of the long time scale process $X_l(t)$.

3.3   Adaptive Bandwidth Allocation Strategy

In this subsection, we describe how to allocate the bandwidth for a system with a buffer of size b and the $M/G/\infty$ input traffic requiring overflow probability less than $e^{-\xi b}$. To allocate the bandwidth for the $M/G/\infty$ input process, we first compute the effective bandwidth C_s of $X_s(t)$ by using Theorem 1. Next, we divide the time axis into *basic allocation periods* of equal size p (slots). At the beginning epoch, say time t, of each basic allocation period we compute the effective bandwidth $\hat{X}_l(t)$ of $X_l(t)$ by using Theorem 2 and (5). Then since the $M/G/\infty$ input traffic $X(t)$ satisfies $X(t) = X_s(t) + X_l(t)$, the bandwidth allocated for the $M/G/\infty$ input traffic is $C_s + \hat{X}_l(t)$ during the current basic allocation period. This procedure is performed continuously for each basic allocation period. By doing this we can capture the short time scale fluctuation by a buffer and the long time scale fluctuation by increasing or decreasing the bandwidth. Furthermore, we only need to estimate the required bandwidth for every period of length p, which reduces the implementation complexity.

4   Numerical Studies

To evaluate the performance of our proposed method, we simulate a number of queueing systems with different parameter values. In simulations since the allocated bandwidth for the $M/G/\infty$ input traffic in our ABA method is $C_s + \hat{X}_l(t)$ for each basic allocation period, the evolution equation of the buffer content process $q(t)$ is given as

$$q(t+1) = \min\left(\max\left(q(t) + X(t) - C_s - \hat{X}_l(t)\ ,\ 0\right)\ ,\ b\right)$$

where $\hat{X}_l(t)$ is updated for each basic allocation period.

To check if the proposed ABA method works well, we consider an $M/G/\infty$ input process with parameters $\lambda = 0.4, \gamma = 1.18$ and $\delta = 0.9153$, and the target overflow probability is given by 10^{-3}. We simulate the queueing process as given above and check the overflow probability of the system. The results for $T = p = 70$ with various values of b are given in Table 1. In the table, we give the mean values and confidence intervals for five sample paths. As seen in the table, the resulting overflow probabilities are very close to our target overflow probability 10^{-3}.

Next, we change the values of T and p to see the effect of T and p on performance. We first fix the value of T and change the value of p, and our experiment show that when p is equal to T, our ABA method performs well. We omit the experiment results due to the limitation of space. So, we propose to use the same value for T and p in our ABA method for simplicity. Now to investigate the effect of T we change the value of T from 10 to 120 and the results are given in Table 2. In this experiment, we use $\lambda = 0.4, \gamma = 1.18, \delta = 0.9153$ for the $M/G/\infty$ input process and the buffer size $b = 150$. As seen in the table, our ABA method performs well when the value of $T(= p)$ is neither small nor

Table 1. Overflow Probability (O.P.) with buffer size b : $\lambda = 0.4$, $\gamma = 1.18$, $\delta = 0.9153$

T value	$T = 70$				
p value	$p = 70$				
b value	$b = 110$	$b = 130$	$b = 150$	$b = 170$	$b = 190$
O.P.	$4.039e^{-4}$	$1.144e^{-3}$	$1.161e^{-3}$	$1.134e^{-3}$	$9.985e^{-4}$
Confidence Interval(C.I.)	$(2.429e^{-4},$ $5.649e^{-4})$	$(8.999e^{-4},$ $1.388e^{-3})$	$(7.896e^{-4},$ $1.533e^{-3})$	$(8.369e^{-4},$ $1.431e^{-3})$	$(5.765e^{-4},$ $1.420e^{-3})$

Table 2. Overflow Probability (O.P.) : $\lambda = 0.4$, $\gamma = 1.18$, $\delta = 0.9153$, $b = 150$

T value	$T = 10$	$T = 30$	$T = 50$	$T = 70$	$T = 100$	$T = 120$
p value	$p = 10$	$p = 30$	$p = 50$	$p = 70$	$p = 100$	$p = 120$
O.P.	0.000	$2.823e^{-5}$	$8.627e^{-4}$	$1.161e^{-3}$	$8.711e^{-4}$	$5.381e^{-5}$
Confidence Interval(C.I.)	$(0.000,$ $0.000)$	$(1.036e^{-5},$ $4.610e^{-5})$	$(6.829e^{-4},$ $1.096e^{-3})$	$(7.896e^{-4},$ $1.533e^{-3})$	$(6.119e^{-4},$ $1.130e^{-3})$	$(1.815e^{-5},$ $8.947e^{-5})$

Table 3. Overflow Probability (O.P.) : $\lambda = 0.4$, $\gamma = 1.5$, $\delta = 0.9153$, $b = 150$

T value	$T = 30$	$T = 50$	$T = 70$	$T = 100$	$T = 120$	$T = 140$
p value	$p = 30$	$p = 50$	$p = 70$	$p = 100$	$p = 120$	$p = 140$
O.P.	0.000	$3.755e^{-5}$	$2.834e^{-4}$	$9.939e^{-4}$	$8.002e^{-4}$	$6.167e^{-5}$
Confidence Interval(C.I.)	$(0.000,$ $0.000)$	$(9.728e^{-6},$ $6.537e^{-5})$	$(8.732e^{-5},$ $4.794e^{-4})$	$(5.853e^{-4},$ $1.403e^{-3})$	$(5.643e^{-4},$ $1.036e^{-3})$	$(6.415e^{-6},$ $1.169e^{-4})$

Table 4. Overflow Probability (O.P.) : $\lambda = 0.4$, $\gamma = 1.9$, $\delta = 0.9153$, $b = 150$

T value	$T = 70$	$T = 100$	$T = 120$	$T = 140$	$T = 160$	$T = 180$
p value	$p = 70$	$p = 100$	$p = 120$	$p = 140$	$p = 160$	$p = 180$
O.P.	$6.541e^{-6}$	$2.707e^{-4}$	$9.360e^{-4}$	$9.722e^{-4}$	$9.999e^{-4}$	$5.698e^{-4}$
Confidence Interval(C.I.)	$(3.064e^{-6},$ $1.002e^{-5})$	$(1.035e^{-4},$ $4.379e^{-4})$	$(6.927e^{-4},$ $1.179e^{-3})$	$(6.587e^{-4},$ $1.286e^{-3})$	$(3.174e^{-4},$ $1.682e^{-3})$	$(9.767e^{-5},$ $1.042e^{-3})$

large. For instance, when T is in the range $[50, 100]$ in our experiment, our ABA method performs well in this case.

Finally, we investigate the effect of the input traffic parameters on performance. To do this, we use $\lambda = 0.4, \delta = 0.9153, b = 150$, but the scale parameter γ is changed from $\gamma = 1.5$ to $\gamma = 1.9$. Note that the long range dependence largely depends on the scale parameter γ, and as the value of γ is getting close to 1, the autocorrelation of the input traffic is getting stronger. The results are given in Tables 3 and 4. As seen in Tables 3 and 4, our ABA method performs well except for very small or large values of T. In addition, we see that when γ is close to 1, our ABA method performs well even under moderately small values of

T. Noting that the effective bandwidth C_s of $X_s(t)$ in Theorem 1 overestimates the required effective bandwidth of $X_s(t)$, when we use large values of T, the use of moderately small values of T is important to implement our ABA method in practice. Hence, we conclude that our ABA method is suitable for the real traffic which exhibits strong correlation.

5   Conclusions

In this paper, we considered a long range dependence traffic which is modelled by an $M/G/\infty$ input process and proposed a new adaptive bandwidth allocation (ABA) method for the long range dependence traffic. In the proposed method, we divide the input process into two sub-processes, a long time scale process and a short time scale process. For the long time scale process we estimate the required bandwidth using the MMSE linear prediction method. On the other hand, for the short time scale process we estimate the effective bandwidth based on the required QoS of the input process. We verified the effectiveness of our proposed method through simulations. Our simulation studies showed that our ABA method is suitable for the real traffic which exhibits strong correlation.

References

1. A. Adas: Supporting real time VBR video using dynamic reservation based on linear prediction. Proc. of IEEE Infocom. (1996) 1476-1483
2. J. Beran, R. Sherman, M. S. Taqqu, W. Willinger: Long-range dependence in variable bit-rate video traffic. IEEE Trans. Commun. **43** (1995) 1566-1579
3. S. Bodamer, J. Charzinski: Evaluation of effective bandwidth schemes for self-similar traffic. Proc. of the 13th ITC Specialist seminar on IP Measurement, Modeling and Management, Monterey, CA, Sep. (2000) 21-1-21-10
4. Cheng-Shang Chang: Performance Guarantees in Communication Networks. Springer (1999) 291-376
5. M. E. Crovella, A. Bestavros: Self-similarity in World Wide Web traffic: Evidence and possible causes. IEEE/ACM Trans. Networking **5** no. 6 (1997) 835-846 ; Proc. ACM SIGMETRICS '96, Philadelphia, May (1996)
6. H. J. Fowler, W. E. Leland: Local area network traffic characteristics with implications for broadband network congestion management. IEEE J. Select. Areas Commun. **9** (1991) 1139-1149
7. M. W. Garrett, W. Willinger: Analysis, modeling, and generation of self-similar VBR video traffic. Proc. SIGCOMM '94 Conf. Sept. (1994) 269-280
8. R. G. Garroppo, S. Giordano, S. Miduri, M. Pagano, F. Russo: Statistical multiplxing of self-similar VBR videoconferencing traffic. Proc. of IEEE Infocom. (1997) 1756-1760
9. A. Karasaridis, D. Hatzinakos: Bandwidth allocation bounds for α-stable self-similar Internet traffic models. IEEE Signal Processing Workshop on Higher-Order Statistics, Ceasarea, Israel, June (1999)
10. M. M. Krunz, A. M. Makowski: Modeling video traffic using $M/G/\infty$ input processes: A compromise between Markovian and LRD Models. IEEE J. Select. Areas Commun. **16** no. 5 (1998) 733-748

11. W. E. Leland, M. S. Taqqu, W. Willinger, D. V. Wilson: On the self-similar nature of Ethernet traffic (extended version). IEEE/ACM Trans. Networking **2** no. 1 Feb. (1994) 1-15
12. A. M. Makowski, M. Parulekar: Buffer asymptotics for $M/G/\infty$ input processes. In: K. Park, W. Willinger (eds.): Self-similar network traffic and performance evaluation. John Wiley & Sons (2002) 215-248
13. M. Parulekar: Buffer engineering for $M/G/\infty$ input processes. Ph.D. dissertation, Univ. Maryland, College Park, Aug. (1999)
14. M. Parulekar, A. M. Makowski: $M/G/\infty$ input processes: a versatile class of models for traffic network. Proc. of IEEE Infocom. (1997) 1452-1459
15. V. Paxson, S. Floyd: Wide area traffic: The failure of Poisson modeling. IEEE/ACM Trans. Networking **3** no. 3 (1993) 226-244
16. F. R. Perlingeiro, L. L. Lee: An effective bandwidth allocation approach for self-similar traffic in a single ATM connection. Proc. of IEEE Globecom. (1999) 1490-1494
17. C. Stathis, B. Maglaris: Modelling the self-similar behavior of network traffic. Computer Networks **34** (2000) 37-47
18. K. P. Tsoukatos, A. M. Makowski: Heavy traffic limits associated with $M/G/\infty$ input processes. Queueing Systems **34** (2000) 101-130

Algorithms for Fast Resilience Analysis in IP Networks *

Michael Menth, Jens Milbrandt, and Frank Lehrieder

University of Würzburg, Institute of Computer Science, Germany
{menth, milbrandt, lehrieder}@informatik.uni-wuerzburg.de

Abstract. When failures occur in IP networks, the traffic is rerouted over the next shortest paths and potentially causes overload on the respective links. This leads to congestion on links and to end-to-end service degradation. These can be anticipated by evaluating the bandwidth requirements of the traffic on the links after rerouting for a set of relevant failure scenarios S. As this set can be large in practice, a fast evaluation of the bandwidth requirements is needed. In this work, we propose several optimized algorithms for that objective together with an experimental assessment of their computation time. In particular, we take advantage of the incremental shortest path first (iSPF) algorithm to reduce the computation time.

1 Introduction

In IP networks, traffic is forwarded according to the shortest path principle. The OSPF or the IS-IS protocol [1, 2] signal topology information through the network by the use of link state advertisements (LSA) such that every node in the network knows all working links with their associated cost metrics. Based on this information, each router can run the shortest path first (SPF) algorithm to calculate the least cost paths and to insert the information about the next hops towards any destination in the network into its routing table. When a node or a link fails, this information is disseminated by the routing protocol to all nodes in the network and the distributed routing tables are recomputed. This is called rerouting and leads to the restoration of traffic forwarding when a failure occurs.

In case of single shortest path (SSP) routing, the router calculates the well-defined next hop for the shortest paths towards any destination (cf. e.g. 7.2.7 in [3]). In case of equal-cost multipath (ECMP) routing, the router identifies all next hops on the paths with equal costs towards any destination and records them in the routing table. As a consequence, the traffic is forwarded almost evenly over all of these next hops. The calculation of the shortest paths is quite time consuming since it scales with $O(n \cdot \log(n))$ with n being the number of nodes in the network. This has a significant impact on the restoration time [4]. If a single link or a single router changes its cost, joins, or disappears, only a small fraction of the shortest paths change. The incremental SPF (iSPF) algorithm recomputes only those shortest paths that are affected by the change. The iSPF algorithm is known from the early days of the ARPANET in the seventies

---

* This work was funded by the Bavarian Ministry of Economic Affairs and the German Research Foundation (DFG). The authors alone are responsible for the content of the paper.

G. Parr, D. Malone, and M. Ó Foghlú (Eds.): IPOM 2006, LNCS 4268, pp. 25–36, 2006.

and has been published in [5]. It speeds up the time for the distributed computation of the shortest paths substantially [6] and it has been implemented recently in routers, e.g., Cisco Systems supports iSPF both for IS-IS and for OSPF [7]. In [8] the complexity regarding comparisons of several iSPF algorithms has been compared experimentally and analytically and another comparison regarding the runtime of Dijkstra's SPF and the iSPF algorithm is provided in [9].

In this paper, we present and assess algorithms to calculate the link utilization for a set of relevant failure scenarios $\mathcal{S}$ in order to detect potential bottlenecks a priori [?, ?]. This set can be rather large even if it contains only all single and double element (link or node) failures. Therefore, the applied algorithms must be fast. They calculate the path layout for all end-to-end (e2e) traffic aggregates in different failure scenarios and based on them the required bandwidth of all links in the network. In contrast to the distributed routing algorithms above, our objective is the computation of the well-defined path layout generated by the distributed calculation of the next hop information. This is achieved by the use of destination graphs. To speed up the computation time, we take a special order of the considered failure scenarios in $\mathcal{S}$ to allow an incremental update of the destination graphs and the required bandwidths. We also adapt the iSPF algorithm to that context to minimize the computation effort. In particular, the iSPF is simplified [10] since it needs to react only to link or node failures but not to new links or nodes, or to the change of their costs. We do not implement any optimization for sorting heaps in the algorithms that can further speed up the calculation [11]. Our results show that our new algorithm is significantly faster than a straightforward naive implementation. Therefore, we recommend their implementation in tools for the a priori detection of overload due to network failures.

The paper is structured as follows. Section 2 presents several optimized algorithms to calculated the bandwidth requirements due to traffic rerouting. Section 3 describes our experimental comparisons of the computation time of these algorithms. In Section 4 we summarize our work and draw our conclusions.

2   Fast Calculation of Resource Requirements in Failure Cases

We represent the network as a graph $\mathcal{G} = (\mathcal{V}, \mathcal{E})$ with $\mathcal{V}$ being the set of nodes and $\mathcal{E}$ being the set of edges. A failure scenario s is characterized by the sets of failed nodes $\hat{\mathcal{V}}(s)$ and links $\hat{\mathcal{E}}(s)$ such that the remaining topology is given by $\mathcal{G}(s) = (\mathcal{V}(s), \mathcal{E}(s))$ with $\mathcal{V}(s) = \mathcal{V} \backslash \hat{\mathcal{V}}(s)$ and $\mathcal{E}(s) = \mathcal{E} \backslash \hat{\mathcal{E}}(s)$. The destination graph $\mathcal{DG}_s^w$ is a directed acyclic graph (DAG) that contains all least cost paths in $\mathcal{G}(s)$ from any source $v \in \mathcal{V}(s)$ to the destination w. To calculate the resource requirements in a special failure case s, we first calculate the destination graphs $\mathcal{DG}_s^w$ for all possible destinations $w \in \mathcal{V}(s)$ and derive then the vector $\hat{\mathbf{r}}_{\mathbf{s}} \in \mathbb{R}_0^{+|\mathcal{E}|}$ of the required bandwidth on all links depending on the choice of single shortest path (SSP) or equal-cost multipath (ECMP) routing [1]. Vectors are printed bold and $\mathbf{0}$ is the vector with only the zero entries.

To construct the destination graphs $\mathcal{DG}_s^w$, we present four different methods with increasing optimization degree.

---

[1] The $|\mathcal{X}|$-operator denotes the cardinality of a set $\mathcal{X}$.

(R0) The simplest one uses Dijkstra's algorithm [12] and recalculates $\hat{\mathbf{r}}_s$ entirely.

(R1) If the destination graphs $\mathcal{DG}_s^w$ for a failure scenario $s \subset s'$ exist, most of the destination graphs $\mathcal{DG}_{s'}^w$ are equal to $\mathcal{DG}_s^w$ and do not need to be computed anew. In this case, the bandwidth vector $\hat{\mathbf{r}}_s$ can be updated incrementally.

(R2Copy) If a destination graph $\mathcal{DG}_{s'}^w$ changes relatively to $\mathcal{DG}_s^w$ due to an additionally failed element, $\mathcal{DG}_{s'}^w$ may be computed based on a copy of $\mathcal{DG}_s^w$ using the iSPF algorithm.

(R2) Copying the full destination graph $\mathcal{DG}_s^w$ may take a long time, therefore, we work with a single copy of $\mathcal{DG}_s^w$ that is reset for another use if it is not needed anymore after its modification to $\mathcal{DG}_{s'}^w$.

2.1   Naive Calculation (R0)

The naive method (R0) calculates the path layout using Dijkstra's algorithm whenever it is needed and computes the bandwidth vector for each scenario $s \in \mathcal{S}$ from scratch.

Basic Path Calculation Based on Dijkstra's Algorithm. Algorithm 1 calculates the destination graph $\mathcal{DG}_s^w$ for a specific destination node w by assigning the appropriate set of predecessor nodes $Pred_s^w(v)$ and successor nodes $Succ_s^w(v)$ to each node $v \in \mathcal{V}(s)$.

The remaining list $\mathcal{R}$ contains all nodes without any successors and the tentative list $\mathcal{T}$ contains all nodes that can still become successor nodes for other nodes and for which shorter paths towards the destination can be possibly found. The function $d(v)$ indicates the distance from any node v to the destination w.

The node $v = argmin_{u \in \mathcal{T}}(d(u))$ has the shortest distance to the destination w among all nodes on the tentative list $\mathcal{T}$ such that its distance cannot be further reduced. Therefore, its set of successors is fixed and v can be removed from the tentative list $\mathcal{T}$. Prior to that, the algorithm checks whether v can be used to find shorter paths to one of its predecessor nodes. This is done by looking at all links l whose destination router $\omega(l)$ is v. The source router of such a link l is denoted by $\alpha(l)$ which is a predecessor node of v within the graph. If a shorter path can be found via v, the set of successors $Succ(\alpha(l))$ is substituted. If routing via v provides an equal-cost path, v is just added to the set $Succ(\alpha(l))$. At last the node v is registered as a predecessor node for all its successors. This information is required to implement the incremental algorithm efficiently. The shortest paths towards the destination w are constructed by calling $Dijkstra(\{w\}, \mathcal{V}(s) \setminus \{w\})$ with the initialization $d(w) = 0$ and $d(v) = \infty$ for $v \neq w$.

Basic Calculation of the Required Bandwidth. We calculate now the required bandwidth of all links in a special failure scenario s based on the destination graphs $\mathcal{DG}_s^w$. The required link bandwidths can be represented by a vector $\hat{\mathbf{r}}_s \in \mathbb{R}_0^{+|\mathcal{E}|}$ whereby each of its entries $\hat{\mathbf{r}}_s(l)$ relates to link l.

The rate of the aggregates from a source v to another destination w is given by the entry $\mathbf{R}(v, w)$ of the traffic matrix $\mathbf{R}$. This rate together with the path of the aggregate that can be derived from the destination graph $\mathcal{DG}_s^w$ induce the aggregate-specific rate vector $\mathbf{r}_s^{v,w} \in \left(\mathbb{R}_0^+\right)^{|\mathcal{E}|}$. Equal-cost multipath (ECMP) routing uses all suitable paths in the destination graph $\mathcal{DG}_s^w$ to forward the traffic, and the traffic is distributed equally over all outgoing interfaces to the destination w. With single shortest path (SSP) routing, the

---

Input: tentative list $\mathcal{T}$, remaining list $\mathcal{R}$
  while $\mathcal{T} \neq \emptyset$ **do**
    $v = argmin_{u \in \mathcal{T}}(d(u))$
    for all $\{l : \omega(l) = v\}$ **do**
      if $d(\alpha(l)) > d(v) + cost(l)$ **then**
        $\{$a shorter path is found for $\alpha(l)\}$
        $d(\alpha(l)) \leftarrow d(v) + cost(l)$
        $Succ(\alpha(l)) \leftarrow \{v\}$
      end if
      if $d(\alpha(l)) = d(v) + cost(l)$ **then**
        $\{$an equal-cost path is found for $\alpha(l)\}$
        $Succ(\alpha(l)) \leftarrow Succ(\alpha(l)) \cup \{v\}$
      end if
      if $\alpha(l) \in \mathcal{R}$ **then** $\{$move visited node to $\mathcal{T}\}$
        $\mathcal{R} \leftarrow \mathcal{R} \setminus \{\alpha(l)\}, \mathcal{T} \leftarrow \mathcal{T} \cup \{\alpha(l)\}$
      end if
    end for
    $\mathcal{T} \leftarrow \mathcal{T} \setminus \{v\}$ $\{$shortest path fixed for $v\}$
    for all $u \in Succ(v)$ **do**
      $Pred(u) \leftarrow Pred(u) \cup \{v\}$
    end for
  end while

Algorithm 1. DIJKSTRA: calculates a unidirectional destination graph $\mathcal{DG}_s^w$ with links towards the destination

traffic is only forwarded towards the next hop with the lowest ID within all equal-cost paths towards the destination w. This is one choice according to 7.2.7 in [3]. Algorithm 2 and Algorithm 3 calculate the aggregate-specific rate vector $\mathbf{r}_s^{v,w}$ for ECMP and SSP. This vector is first initialized by $\mathbf{r}_s^{v,w} = \mathbf{0}$ before the algorithms are called by RATEVECTORX($\mathbf{r}_s^{v,w}, v, w, \mathbf{R}(v,w)$) with $X \in \{ECMP, SSP\}$. In case of ECMP, Algorithm 2 distributes the rate c at the node v over the links towards the successors $Succ_s^w(v)$ within the destination graph $\mathcal{DG}_s^w$. In case of SSP, Algorithm 3 distributes the traffic from any node v over the single link towards its successor node with the lowest node ID in the destination graph $\mathcal{DG}_s^w$. If a node v has failed, the destination graph $\mathcal{DG}_s^v$ does not exist and no other destination graph $\mathcal{DG}_s^x$ contains v. Thus, the aggregate-specific rate vectors $\mathbf{r}_s^{x,v} = \mathbf{0}$ and $\mathbf{r}^{v,x} = \mathbf{0}$ are zero.

The vector of the required link bandwidth $\hat{\mathbf{r}}_s$ is computed as the sum of all aggregate-specific rate vectors

$$\hat{\mathbf{r}}_s = \sum_{v,w \in \mathcal{V}: v \neq w} \mathbf{r}_s^{v,w}. \tag{1}$$

2.2 Incremental Naive Calculation (R1)

The incremental naive method takes advantage of the fact that the set of protected failure scenarios $\mathcal{S}$ contains many similar failure scenarios s' being a superset of others

Input: destination graph $\mathcal{DG}_s^w$, rate vector $\mathbf{r}$, node v, destination w, rate c

$c' \leftarrow \frac{c}{|Succ_s^w(v)|}$

    for all $u \in Succ_s^w(v)$ **do**

        $\mathbf{r}(l(v,u)) \leftarrow \mathbf{r}(l(v,u)) + c'$

        if $u \neq w$ **then**

            $RateVectorECMP(\mathcal{DG}_s^w, \mathbf{r}, u, w, c')$

        end if

    end for

Algorithm 2. RATEVECTORECMP: calculates the aggregate-specific rate vector $\mathbf{r}$ induced by a flow from v to w with rate c for ECMP routing

Input: destination graph $\mathcal{DG}_s^w$, rate vector $\mathbf{r}$, node v, destination w, rate c

    $u \leftarrow argmin_{u' \in Succ_s^w(v)}(ID(u'))$

    $\mathbf{r}(l(v,u)) \leftarrow \mathbf{r}(l(v,u)) + c$

    if $u \neq w$ **then**

        $RateVectorSSP(\mathcal{DG}_s^w, \mathbf{r}, u, w, c)$

    end if

Algorithm 3. RATEVECTORSSP: calculates the aggregate-specific rate vector $\mathbf{r}$ induced by a flow from v to w with rate c for ECMP routing

$(s \subseteq s')$. It saves computation time for the failure-specific destination graphs $\mathcal{DG}_s^w$ and allows an incremental calculation of the bandwidth vector $\hat{\mathbf{r}}_s$.

Selective Path Calculation Using Dijkstra's Algorithm. If two failure scenario s and s' are similar, most of their destination graphs $\mathcal{DG}_s^w$ and $\mathcal{DG}_{s'}^w$ do not differ. In particular, if s is a subset of s' ($s \subset s'$), $\mathcal{DG}_{s'}^w$ only differs from $\mathcal{DG}_s^w$ if $\mathcal{DG}_s^w$ contains an element of the set difference $\Delta_s = s' \setminus s$. Thus, we construct a function $Contains_s^w(x)$ whenever we build a new destination graph $\mathcal{DG}_s^w$. In addition, we take advantage of the $Contains$-function and arrange all failure scenarios $s \in S$ hierarchically in such a way that that supersets s' are subordinate to subsets s. An example for such a hierarchy is given in Figure 1. As a result, if $(s \subset s')$ holds, the destination graph $\mathcal{DG}_{s'}^w$ can be overtaken from $\mathcal{DG}_s^w$ unless $Contains_s^w(x)$ is *true* for any $x \in \Delta_s$.

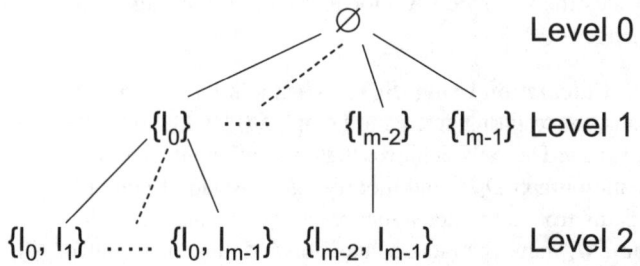

Fig. 1. The failure scenarios $s \in S$ are organized in a tree structure such that each failure scenario is a child of one of its subsets

Incremental Calculation of the Required Link Bandwidth $\hat{r}_{s'}$ Based on Recalculated Destination Graphs $\mathcal{DG}_s^w$. If the path of the aggregate is the same in the failure scenario s and s', $r_s^{v,w}$ can be used instead of $r_{s'}^{v,w}$ for the calculation of the required bandwidth vector in Equation (1). The path is the same if s contains only a subset of the failures in s' and if all links and nodes in $\mathcal{DG}_s^w$ are working in $\mathcal{DG}_{s'}^w$, too. Algorithm 4 calculates the vector of the required link bandwidths for failure scenario s' based on the one for s. It uses the function $Contains_s^w(x)$ to find all aggregates whose destination graph $\mathcal{DG}_{s'}^w$ has changed with respect to $\mathcal{DG}_s^w$, calculates for these aggregates the new aggregate-specific rate vector $r_{s'}^{v,w}$, and updates the vector for the required link bandwidths $\hat{r}_{s'}$ incrementally.

Input: required bandwidth vector $\hat{r}_s$, failure scenarios s and s'

    $\hat{r}_{s'} \leftarrow \hat{r}_s, \Delta_s \leftarrow s' \setminus s$

    for all $w \in \mathcal{V}$ **do**

        $equal \leftarrow true$

        for all $x \in \Delta_s$ **do**

            if $Contains_s^w(x)$ **then**

                $equal \leftarrow false$

            end if

        end for

        if $equal \neq true$ **then**

            for all $v \in \mathcal{V}$ **do**

                $r_{s'}^{v,w} = 0$

                RATEVECTORX$(\mathcal{DG}_s^w, r_{s'}^{v,w}, v, w, \mathbf{R}(v,w))$

                $\hat{r}_{s'} \leftarrow \hat{r}_{s'} - r_s^{v,w} + r_{s'}^{v,w}$

            end for

        end if

    end for

Output: required bandwidth vector $\hat{r}_{s'}$

Algorithm 4. INCREMENTALREQUIREDBANDWIDTH: calculates the required bandwidth vector $\hat{r}_{s'}$ based on $\hat{r}_s$ incrementally

2.3   Incremental Calculation Based on iSPF (R2Copy)

The incremental method based on iSPF has two advantages compared to the incremental naive calculation: it requires less effort to construct a new required destination graph $\mathcal{DG}_s^w$ and updates the bandwidth vector $\hat{r}_s$ by only those aggregates whose path has effectively changed.

Selective Path Calculation Using iSPF. When a link l or node v within a destination graph $\mathcal{DG}_s^w$ fails, the resulting destination graph $\mathcal{DG}_{s'}^w$ with updated shortest paths must be constructed anew. The iSPF achieves that in an efficient way by copying the existing, similar destination graph $\mathcal{DG}_s^w$ and modifying it instead of computing it entirely from scratch. The paths from all nodes v that reach the destination node w only via a failed link or node in $\mathcal{DG}_s^w$ have then lost connection and must be rerouted. Afterwards, only all nodes v that contain the failed network element in their paths towards the destination

---

Input: failed link l, set of disconnected network elements $\mathcal{R}$
   $Pred(\omega(l)) \leftarrow Pred(\omega(l)) \setminus \{\alpha(l)\}$
   $Succ(\alpha(l)) \leftarrow Succ(\alpha(l)) \setminus \{\omega(l)\}$
   if $|Succ(v)| = 0$ **then** $\{v \text{ disconnected}\}$
      $d(v) \leftarrow \infty, \mathcal{R} \leftarrow \mathcal{R} \cup \{v\}$
      for all $u \in Pred(v)$ **do**
         $RemoveLink(l(u,v), \mathcal{R})$
      end for
   end if

Algorithm 5. REMOVELINK: removes from the destination graph $\mathcal{DG}_s^w$ all network elements that have lost connection due to the failure of link l

---

Input: failed node v, set of disconnected network elements $\mathcal{R}$
   $\mathcal{R} \leftarrow \mathcal{R} \cup \{v\}$,
   $d(v) \leftarrow \infty$
   for all $u \in Succ(v)$ **do**
      $Pred(u) \leftarrow Pred(u) \setminus \{v\}, Succ(v) \leftarrow Succ(v) \setminus \{u\}$
   end for
   for all $u \in Pred(v)$ **do**
      $RemoveLink(l(u,v), \mathcal{R})$
   end for

Algorithm 6. REMOVENODE: removes from the destination graph $\mathcal{DG}_s^w$ all network elements that have lost connection due to the failure of node v

w need to recompute their rate vectors $\mathbf{r}_{s'}^{v,w}$ using the algorithms RATEVECTORX in Section 2.1; the other rate vectors $\mathbf{r}_{s'}^{v,w}$ are equal to $\mathbf{r}_s^{v,w}$.

The recursive Algorithm 5 removes from the destination graph $\mathcal{DG}_s^w$ all network elements that have lost connection due to the failure of link l and adds the disconnected nodes to the set $\mathcal{R}$.

Algorithm 6 removes a node v from the destination graph by disconnecting it explicitly from all its successor nodes and by disconnecting it implicitly from all its predecessor nodes by calling REMOVELINK$(l, \mathcal{R})$ for all links leading to v[2]. The failed node v is not added to the set of disconnected nodes like in Algorithm 5 since it should not be reconnected to the graph.

Algorithm 7 reconnects the disconnected working nodes in $\mathcal{R}$ by first connecting them to the connected structure of the remaining destination graph $\mathcal{DG}_s^w$ and moving then the freshly connected nodes to the tentative list $\mathcal{T}$. Finally, DIJKSTRA$(\mathcal{T}, \mathcal{R})$ is called and completes the destination graph $\mathcal{DG}_{s'}^w$.

Incremental Calculation of the Required Link Bandwidth $\hat{r}_{s'}$ Based on Recalculated Destination Graphs $\mathcal{DG}_s^w$. The iSPF limits the overhead to reroute paths that are affected by a link or a node failure. In addition, the incremental update of the

---

[2] The link from node u to node v is denoted by $l(u,v)$.

Input: set of disconnected working nodes $\mathcal{R}$
  for all $v \in \mathcal{R}$ **do**
    for all $\{l : \alpha(l) = v\}$ **do**
      if $d(\omega(l)) < \infty$ **then** $\{\omega(l)$ has a path to $w\}$
        if $d(\omega(l)) + cost(l) < d(v)$ **then**
          $\{$a shorter path from v to w is found$\}$
          if $d(v) = \infty$ **then**
            $\{$move v from remaining to tentative list$\}$
            $\mathcal{R} \leftarrow \mathcal{R} \setminus \{v\}, \mathcal{T} \leftarrow \mathcal{T} \cup \{v\}$
          end if
          $d(v) \leftarrow d(\omega(l)) + cost(l), Succ(v) \leftarrow \{w\}$
        else if $d(\omega(l)) + cost(l) = d(v)$ **then**
          $\{$an equal-cost path to v is found$\}$
          $Succ(v) \leftarrow Succ(v) \cup \{w\}$
        end if
      end if
    end for
  end for
  $Dijkstra(\mathcal{T}, \mathcal{R})$

Algorithm 7. RECONNECTNODES: reconnects the disconnected working nodes in $\mathcal{R}$ to the destination graph $\mathcal{DG}_s^w$ and creates thereby $\mathcal{DG}_{s'}^w$

required link bandwidth can be limited to those nodes within a destination graph whose ECMP paths have changed. We find them by identifying the indirect predecessor nodes of a failed link or node within the base destination graph $\mathcal{DG}_s^w$. Algorithm 8 collects all predecessor nodes of the node v recursively and stores them in the set $\mathcal{C}$. At the beginning of the algorithm, the set of collected nodes is empty, i.e. $\mathcal{C} = \emptyset$. If a node v fails, we collect COLLECTINDIRECTPREDECESSOR$(v, \mathcal{C})$ and if a link l fails, we collect COLLECTINDIRECTPREDECESSOR$(\alpha(l), \mathcal{C})$. Finally, the set $\mathcal{C}$ contains all nodes that have a changed path layout in $\mathcal{DG}_{s'}^w$ compared to $\mathcal{DG}_s^w$. As a consequence, the incremental update of the bandwidth vector $\hat{r}_s$ in Algorithm 4 can be limited to the nodes in $\mathcal{C}$.

Input: node v, set of indirect predecessors $\mathcal{C}$
  for all $u \in Pred(v)$ **do**
    if $u \notin \mathcal{C}$ **then**
      $\mathcal{C} \leftarrow \mathcal{C} \cup \{u\}$
      COLLECTINDIRECTPREDECESSORS$(u, \mathcal{C})$
    end if
  end for

Algorithm 8. COLLECTINDIRECTPREDECESSORS: collects in the set $\mathcal{C}$ all indirect predecessor nodes of node v

2.4   Incremental Calculation Based on iSPF with Reduced Copy Overhead (R2)

We discuss some implementation issues regarding an efficient memory management which finally leads to the improved version R2 with respect to R2Copy.

When the bandwidth requirements $r_s^{v,w}$ of many failure scenarios are computed, many destination graphs $\mathcal{DG}_s^w$ are sequentially constructed and evaluated. Deleting such a graph after its analysis and constructing a new, similar one requires quite an effort for memory allocation, which should be avoided if possible. The naive calculation in Section 2.1 recomputes all graphs from scratch.

To make this more efficient, the sets $Pred(v)$ and $Succ(v)$ of the old destination graph $\mathcal{DG}_s^w$ may be emptied and the new destination graph $\mathcal{DG}_{s'}^w$ may be constructed reusing the nodes from the old destination graph $\mathcal{DG}_s^w$.

The incremental naive calculation in Section 2.2 recomputes only those graphs $\mathcal{DG}_{s'}^w$ from scratch that have changed with regard to a predecessor destination graph $\mathcal{DG}_s^w$. The overall analysis traverses all failure scenarios of interest $\mathcal{S}$ recursively along a tree structure (cf. Figure 1) Thus, the destination graph for a specific destination w may change for each of the failure scenarios $s_0 \subset s_1 \subset ... \subset s_n$. Therefore, a complete set of nodes must be available on each level of the tree to construct the destination graph.

The incremental calculation based on the iSPF algorithm in Section 2.3 requires not only a new set of nodes but a copy of the destination graph $\mathcal{DG}_s^w$ that serves as a base for the construction of the destination graph $\mathcal{DG}_{s'}^w$ using iSPF. When $\mathcal{DG}_{s'}^w$ is not needed anymore, only the sets $Pred(v)$ and $Succ(v)$ of those nodes need to be reset that have been changed relative to the one in $\mathcal{DG}_s^w$. This saves the entire deletion of the current connectivity of $\mathcal{DG}_{s'}^w$ and generating a new copy of $\mathcal{DG}_s^w$. We call this method R2.

3   Comparison of Experimental Computation Times

We implemented the above presented algorithms in Java 1.5.0_06 and tested the computation time experimentally on a standard PC Pentium M, 1.86 GHz with 1 GB RAM and WinXP Pro SP2. We use random topologies in our study for which the most important network characteristics are the network size in terms of nodes $n = |\mathcal{V}|$ and links $m = |\mathcal{E}|$. They define the average node degree $deg_{avg} = \frac{2 \cdot m}{n}$ which indicates the average number of adjacent links of a node and which is thereby an indirect measure for the network connectivity. We use the topology generator from Section 4.4.2 in [13] to control the minimum and the maximum node degree deg_{min} and deg_{max} which are limited by the maximum deviation deg_{dev}^{max} of the node degrees from their average value. It generates connected networks and avoids loops and parallels.

3.1   Comparison of Computation Times

We consider networks of different sizes with an average node degree $deg_{avg} \in \{3, 4, 5, 6\}$ and a maximum deviation from the average node degree of $deg_{dev}^{max} \in \{1, 2, 3\}$. We randomly generate 5 networks of each combination. Figures 2(a) and 2(b) show the time for the computation of the ECMP routing and the link load for failures of single network elements and for failures of up to two network elements, respectively. The computation time is given in seconds for the naive method (R0) depending on the network size in nodes. The x-axes of both figures have a different scale since the calculation of the double failure scenarios is very time-consuming. We fit the experimental computation time of R0 by a function of the form $O(n^k)$ and derive k from an approximation that minimizes the sum of the squared deviations from the experimental results. In the single failure case, the experimental computation time grows approximately like $O(n^{3.36})$ (dashed line) with the number of nodes n in the network which results from a $O(n^2)$ worst case runtime of the Dijkstra algorithm and an $O(n)$ number of considered failure scenarios $(\binom{n+m}{0} + \binom{n+m}{1})$. In the double failure case, we observe a growths of about $O(n^{5.44})$ which is due to a larger number of failure scenarios $(\binom{n+m}{0} + \binom{n+m}{1} + \binom{n+m}{2})$. This is the practical runtime of the program for small network instances and all software overhead while the theoretical runtime of the mere algorithm is bounded by $O(n^4)$.

The computation time for the incremental naive method is presented relative to the one for R0. Surprisingly, the incremental naive method (R1) takes about the same time as the naive method (R0). Data structures for the implementation of the function $Contains_s^w(x)$ must be updated whenever the destination graph $\mathcal{DG}_s^w$ is reconstructed by R1. This makes the algorithm more complex. In addition, the destination graphs contain often more than the minimum number of $n-1$ links since equal-cost paths frequently occur due to our hop metric assumption, and must be updated in more than $\frac{n-1}{m}$ of all cases. As a consequence, the savings of destination graph calculations of R1 are too small to achieve a considerable speedup for its computation time compared to the one of R0. This is different for the incremental calculation based on iSPF (R2) which requires only 10% of the computation time of R0. This holds only, if the data structures are reused. If the data structures are copied (R2Copy), we still see significant savings of up to 75%, but compared to (R2), the computation time takes four times longer in large networks. The confidence intervals in both figures are based on a confidence level of 95% to guarantee that the results from our experiments are sufficiently accurate.

3.2   Sensitivity Analysis Regarding Network Connectivity

To underline the above observations, we conduct a sensitivity analysis of the computation time regarding the average node degree deg_{avg} of the networks. Figures 3(a)

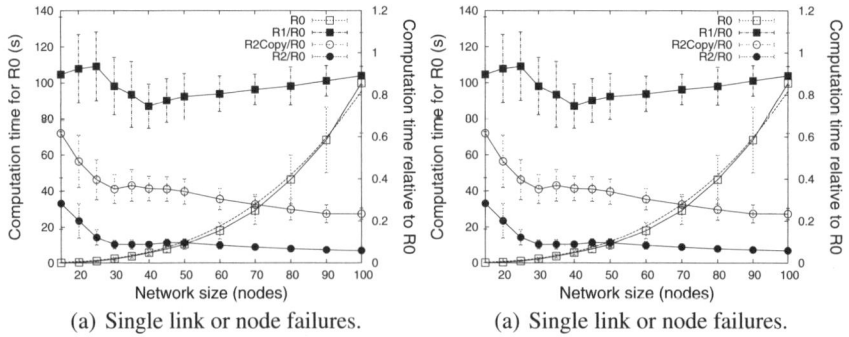

(a) Single link or node failures.   (a) Single link or node failures.

Fig. 2. Comparison of the computation time of the naive calculation (R0), the incremental naive calculation (R1), the incremental calculation based on iSPF with (R2) and without copy reduction (R2Copy)

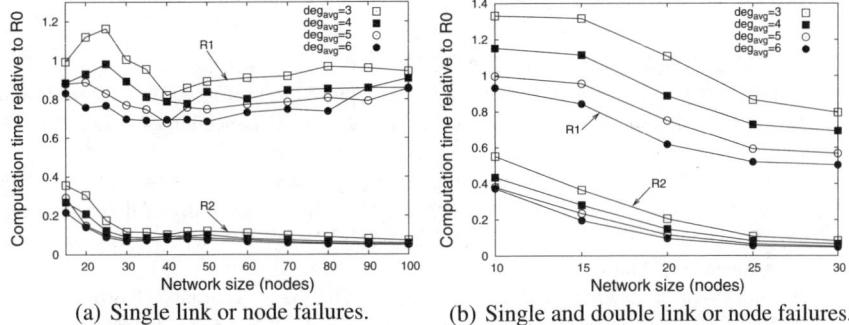

(a) Single link or node failures.   (b) Single and double link or node failures.

Fig. 3. Comparison of the computation time of the incremental naive calculation (R1) and the incremental calculation based on iSPF with copy reduction (R2) depending on the node degree of the networks

and 3(b) show the relative computation time of R1 and R2 compared to R0 separately for networks with different node degrees. The curves for both R1 and R2 show that networks with a large node degree like $deg_{avg} = 6$ lead to larger time savings than networks with small node degrees like $deg_{avg} = 3$. Networks with a large average node degree have more links than those with a small one, but their destination graphs contain approximately the same number of links since $(n-1)$ links already form a spanning tree. As a consequence, in networks with the same number of nodes but a larger number of links it is less likely that a destination graph is affected by a link failure. Thus, they offer an increased savings potential for destination graph calculations. However, if the average node degree and the network size are small, the optimization method R1 can lead to clearly increased computation time and becomes counterproductive. These findings are very well visible if up to two network elements fail. For R2 we observe basically the same phenomenon, but its computation time is mostly limited to 20% or less of the one for R0. Hence, the proposed method R2 effectively reduces the computation time for programs that analyze the network resilience.

4   Conclusion

In this work we have presented a simple and a complex optimized method to speed up the calculation of the (re-)routing and the link load in a network for a large set of different failure scenarios. The reference model for our performance comparison is Dijkstra's shortest path first algorithm (R0). The simple method (R1) just skips the recalculation of a destination graph if it does not contain the failed network element. However, this achieves hardly any speedup. The complex method (R2) is based on an incremental shortest path first calculation and on a careful reuse strategy for data structures. It reduces the computation time to 10% while without the reuse strategy, the computation time is decreased to 25%. Hence, computer programs for the analysis of the network resilience should implement the complex method with a careful reuse strategy for data structures as it considerably accelerates the calculation of the routing and the traffic distribution for a large set of failure cases.

References

1. J. Moy, "RFC2328: OSPF Version 2," April 1998.
2. ISO, "ISO 10589: Intermediate System to Intermediate System Routing Exchange Protocol for Use in Conjunction with the Protocol for Providing the Connectionless-Mode Network Service," 1992.
3. D. Oran, "RFC1142: OSI IS-IS Intra-Domain Routing Protocol," Feb. 1990.
4. G. Iannaccone, C.-N. Chuah, S. Bhattacharyya, and C. Diot, "Feasibility of IP Restoration in a Tier-1 Backbone," *IEEE Network Magazine (Special Issue on Protection, Restoration and Disaster Recovery)*, March 2004.
5. J. M. McQuilan, I. Richer, and E. C. Rosen, "The New Routing Algorithm for the ARPANET," *IEEE Transactions on Communications*, vol. 28, no. 5, May 1980.
6. P. Francois, C. Filsfils, J. Evans, and O. Bonaventure, "Achieving Sub-Second IGP Convergence in Large IP Networks," *ACM SIGCOMM Computer Communications Review*, vol. 35, no. 2, pp. 35 – 44, July 2005.
7. J.-P. Vasseur, M. Pickavet, and P. Demeester, *Network Recovery*, 1st ed.   Morgan Kaufmann / Elsevier, 2004.
8. P. Narvaez, "Routing Reconfiguration in IP Networks," PhD thesis, Massachusetts Institut of Technology (MIT), June 2000.
9. H. El-Sayed, M. Ahmed, M. Jaseemuddin, and D. Petriu, "A Framework for Performance Characterization and Enhancement of the OSPF Routing Protocol," in *IASTED International Conference on Internet and Multimedia Systems and Applications (EuroIMSA)*, Grindelwald, Switzerland, Feb. 2005.
10. S. Nelakuditi, S. Lee, Y. Yu, and Z.-L. Zhang, "Failure Insensitive Routing for Ensuring Service Availability," in *IEEE International Workshop on Quality of Service (IWQoS)*, 2003.
11. L. Buriol, M. Resende, and M. Thorup, "Speeding up Dynamic Shortest Path Algorithms," AT&T Labs Research, Technical Report TD-5RJ8B, 2003.
12. E. W. Disjkstra, "A Note on Two Problems in Connexion with Graphs," *Numerische Mathematik*, vol. 1, pp. 269 – 271, 1959.
13. M. Menth, "Efficient Admission Control and Routing in Resilient Communication Networks," PhD thesis, University of Würzburg, Faculty of Computer Science, Am Hubland, July 2004.

Efficient OSPF Weight Allocation for Intra-domain QoS Optimization[*]

Pedro Sousa[1], Miguel Rocha[1], Miguel Rio[2], and Paulo Cortez[3]

[1] Dep. of Informatics, University of Minho, Portugal
{pns, mrocha}@di.uminho.pt
[2] Dep. of Electronic and Electrical Engineering, University College London, UK
m.rio@ee.ucl.ac.uk
[3] Dep. of Information Systems, University of Minho, Portugal
pcortez@dsi.uminho.pt

Abstract. This paper presents a traffic engineering framework able to optimize OSPF weight setting administrative procedures. Using the proposed framework, enhanced *OSPF* configurations are now provided to network administrators in order to effectively improve the QoS performance of the corresponding network domain. The envisaged NP-hard optimization problem is faced resorting to Evolutionary Algorithms, which allocate OSPF weights guided by a bi-objective function. The results presented in this work show that the proposed optimization tool clearly outperforms common weight setting heuristics.

1 Introduction

The onset of new types of applications and their incremental integration in IP based networks have fostered the development of novel network solutions, aiming at providing end-users with Quality of Service (QoS) support [1]. In the context of a QoS aware network domain, Internet Service Providers (ISPs) have Service Level Agreements (SLAs) [2] with their clients and with peered ISPs that have to be strictly obeyed in order to avoid financial penalties. To successfully face such requirements, there is an important set of configuration tasks that have to be performed by administrators in order to assure that correct resource provisioning is achieved in the ISP domain.

Independently of the large set of mechanisms and alternatives that might be in place in any QoS capable infrastructure, there are some components which, by their nature, have crucial importance irrespective of the particular QoS solution adopted. One of such components has the ability to control the data path followed by packets traversing a given Wide Area Network (WAN). In a TCP/IP WAN, consisting of a single administrative domain, there are alternative strategies for this purpose: Intra-domain routing protocols or Multi-Protocol Label Switching (MPLS) [3]. However, the use of MPLS presents some drawbacks when used in the context of packet switching, when compared

[*] The authors wish to thank the Portuguese National Conference of Rectors (CRUP)/British Council Portugal (B-53/05 grant), the Nuffield Foundation (NAL/001136/A grant), the Engineering and Physical Sciences Research Council (EP/522885 grant) and Project SeARCH (Services and Advanced Research Computing with HTC/HPC clusters), funded by FCT.

G. Parr, D. Malone, and M. Ó Foghlú (Eds.): IPOM 2006, LNCS 4268, pp. 37–48, 2006.

with the simplicity of some routing protocols. As regards intra-domain routing proto-cols, the most commonly used today is Open Shortest Path First (OSPF) [4][5]. Since, in OSPF, the link weight setting process is the only way administrators can affect the network behavior, this choice is of crucial importance. Nevertheless, in practice, simple rules of thumb are typically used in this task, like setting the weights inversely propor-tional to the link capacity. This approach often leads to sub-optimal network resource utilization. An ideal way to improve the process of OSPF weight setting is to imple-ment traffic engineering. This was the approach taken by Fortz et al [6] where this task was viewed as an optimization problem by defining a cost function that measures the network congestion. The same authors proved that this task is a NP-hard problem and proposed some local search heuristics that compared well with the MPLS model. How-ever, such approach did not accommodate delay based constraints that are also crucial to implement QoS aware networking services in the Internet.

In this paper, Evolutionary Algorithms (EAs) are employed to calculate link-state routing weights, that optimize traffic congestion, while simultaneously complying to specific delay requirements. In this way, the framework proposed in this paper should be viewed as a network management tool which, while focusing only at the OSPF rout-ing level, aims at optimizing the overall QoS performance of a given domain. The paper is organized as follows: firstly, the problem is defined and the EAs designed to tackle this problem are described; the following section presents the experiments and corre-sponding results; finally, conclusions are drawn and the future work is revealed.

2   Problem Formulation

The network scenario of Fig. 1(a) includes a set of network nodes which are intercon-nected using links with distinct capacities and propagation delays. It is assumed that the ISP can map the clients demands into a matrix (there are several techniques on how to obtain such matrices, e.g. Medina et al [7]). This matrix summarizes, for each source/destination router pair, a given amount of bandwidth and end-to-end delay re-quired to be supported by the ISP. Figure 1(a) shows a scenario involving an individual demand between the network nodes A and B. Assuming that this demand is mainly expressed in terms of a delay target, then the ISP, in the absence of other traffic, should be able to compute OSPF weights that will result in a data path with the minimum end-to-end delay between A and B (see PATH 2 in Fig. 1(a)). In opposition, if no delay requirements are imposed in the demand, and the only constraint between A and B is a given bandwidth requirement (e.g. 90Mbps), then the optimization methods would try to minimize the network congestion and assign OSPF weights to force a data path inducing the lowest level of losses in the traffic (PATH 1 in Fig. 1(a)). For these two dis-tinct optimization aims, two distinct sets[1] of OSPF weights are presented in Fig. 1(b)(c). The OSPF weights are assigned in order that, after running the Dijkstra algorithm, the shortest paths between nodes A and B are a perfect match of PATH 1 and PATH 2.

Additionally, note that in Fig. 1(a) if a given demand has simultaneously bandwidth and delay constraints, it is expected that the OSPF weights set by the optimization algorithms are chosen in order to find a data path representing a tradeoff between the

---

[1] To simplify, in the selected examples the OSPF weights range from 1 to 3.

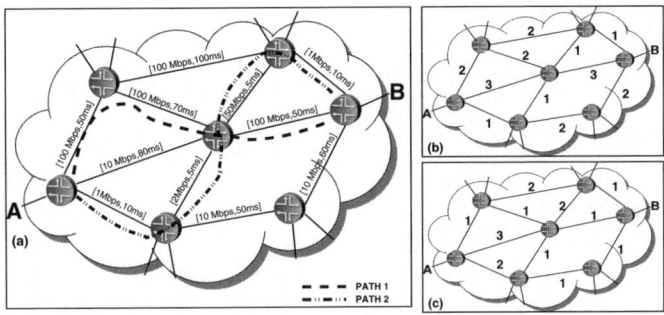

Fig. 1. Example of a network scenario with distinct end-to-end paths between nodes A and B

bandwidth and delay metrics. The example of Fig. 1(a) is extremely simple, due to the fact that one simple demand was considered in the traffic traversing the network domain. Considering now that each router pair of a given ISP has specific bandwidth and delay demands, it is easy to understand how difficult it is to correctly set OSPF weights using simple heuristics. Although this work assumes OSPF scenarios with an unique level of weights, which might be considered as more challenging and difficult for the optimization purposes, it is also intended to adapt the proposed optimization model for OSPF schemes considering multiple levels of weights.

The general routing problem [8], that underpins our work, represents routers and transmission links by a set of nodes (N) and a set of arcs (A) in a directed graph $G = (N, A)$. In this model, c_a represents the capacity of each link $a \in A$. Additionally, a demand matrix D is available, where each element d_{st} represents the traffic demand between each pair of nodes s and t from N. Let us assume that, for each arc a, the variable $f_a^{(st)}$ represents how much of the traffic demand between s and t travels over arc a. The total load on each arc a (l_a) can be defined by Eq. (1), while the link utilization rate u_a is given by Eq. (2). It is then possible to define a congestion measure for each link (Φ_a), using a cost function p that has small penalties for values near 0. However, as the values approach the unity it becomes more expensive and exponentially penalizes values above 1 (Fig. 2). Given this function, the congestion measure for a given arc can be defined by Eq. (3). Under this framework, it is possible to define a linear programming instance, where the purpose is to set the value of the variables f_a^{st} that minimize the objective function defined by Eq. (4). The complete formulation can be found in [6]. In the following the optimal solution to this problem is denoted by Φ_{Opt}.

$$l_a = \sum_{(s,t) \in N \times N} f_a^{st} \ (1) \qquad u_a = \frac{l_a}{c_a} \ (2) \qquad \Phi_a = p(u_a) \ (3) \qquad \Phi = \sum_{a \in A} \Phi_a \ (4)$$

In OSPF, all arcs are associated with an integer weight. All nodes use these weights in the Dijkstra algorithm [9] to calculate the shortest paths to all other nodes. Each of these paths has a length equal to the sum of its arcs. All the traffic from a given source to a destination travels along the shortest path. If there are two or more paths with the same length, between a given source and a destination, traffic is evenly divided among the arcs in these paths (load balancing) [10]. Let us assume a given solution,

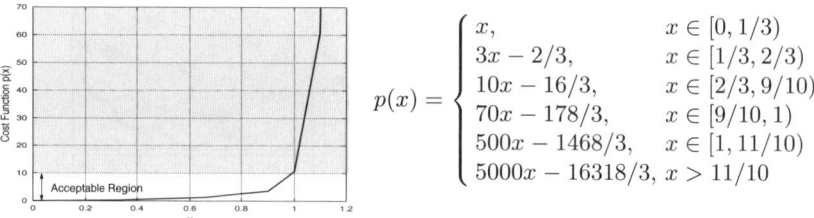

$$p(x) = \begin{cases} x, & x \in [0, 1/3) \\ 3x - 2/3, & x \in [1/3, 2/3) \\ 10x - 16/3, & x \in [2/3, 9/10) \\ 70x - 178/3, & x \in [9/10, 1) \\ 500x - 1468/3, & x \in [1, 11/10) \\ 5000x - 16318/3, & x > 11/10 \end{cases}$$

Fig. 2. Representation and definition of the penalty function $p(x)$

i.e. a weight assignment (w), and the corresponding utilization rates on each arc(u_a). In this case, the total routing cost is expressed by Eq. (5), for the loads and corresponding penalties ($\Phi_a(w)$) calculated based on the given OSPF weights. In this way, the OSPF weight setting problem (as defined in [6]) is equivalent to finding the optimal weight values for each link (w_{opt}), in order to minimize the function $\Phi(w)$. The congestion measure can be normalized over distinct topology scenarios, by using a scaling factor defined in [6] (Eq. (6)), where h_{st} is the minimum hop count between nodes s and t.

$$\Phi(w) = \sum_{a \in A} \Phi_a(w) \quad (5) \qquad \Phi_{UNCAP} = \sum_{(s,t) \in N \times N} d_{st} h_{st} \quad (6)$$

Finally, the scaled congestion measure cost is defined as Eq. (7) and the relationships defined in Eq. (8) hold, where $\Phi OptOSPF^*$ is the normalized congestion imposed by the optimal solution to the OSPF weight setting problem. It is important to note that when Φ^* equals 1, all loads are below $1/3$ of the link capacity; on the other hand, when all arcs are exactly full the value of Φ^* is 10 2/3. This value will be considered as a threshold that bounds the acceptable working region of the network.

$$\Phi^*(w) = \frac{\Phi(w)}{\Phi_{UNCAP}} \quad (7) \qquad 1 \leq \Phi^*_{OPT} \leq \Phi^*_{OptOSPF} \leq 5000 \quad (8)$$

Delay requirements were modeled as a matrix DR, that for each pair of nodes $(s,t) \in N \times N$ (where $d_{st} > 0$) gives the delay target for traffic between s and t (denoted by DR_{st}). In a way similar to the congestion model presented before, a cost function was developed to evaluate the delay compliance for each scenario. This function takes into account the average delay of the traffic between the two nodes (Del_{st}), a value calculated by considering all paths between s and t with minimum cost and averaging the delays in each. The delay in each path is the sum of the propagation delays in its arcs ($Del_{st,p}$) and queuing delays in the nodes along the path ($Del_{st,q}$). Note that in some network scenarios the latter component might be neglected (e.g. if the propagation delay component has an higher order of magnitude than queuing delays). However, if required, the $Del_{st,q}$ component might be approximated, resorting to queueing theory [11], taking into account the following parameters at each node: the capacity of the corresponding output link (c_a), the link utilization rate (l_a) and more specific parameters such as the mean packet size and the overall queue size associated with the link. The delay compliance ratio for a given pair $(s,t) \in N \times N$ is, therefore, defined by Eq. (9). As before, a penalty for delay compliance can be calculated using function p. So, the γ_{st} function is defined according to Eq. (10). This, in turn, allows the definition of a delay minimization cost function, for a given a set of OSPF weights (w), expressed by Eq. (11). In Eq. (11), the $\gamma_{st}(w)$ values represent the delay penalties for each end-to-end path, given the routes determined by the OSPF weight set w. This function can be normalized dividing the values by the sum of all minimum end-to-end delays[2], as expressed by Eq. (12). It is now possible to define the optimization problem addressed in this work. Indeed, given a network represented by a graph G of nodes N and arcs A, a demand matrix D and a delay requirements matrix DR, the aim is to find the set of OSPF weights that simultaneously minimizes the functions $\Phi^*(w)$ and $\gamma^*(w)$.

$$dc_{st} = \frac{Del_{st}}{DR_{st}} \qquad (9)$$

$$\gamma_{st} = p(dc_{st}) \qquad (10)$$

$$\gamma(w) = \sum_{(s,t) \in N \times N} \gamma_{st}(w) \qquad (11)$$

$$\gamma^*(w) = \frac{\gamma(w)}{\sum_{(s,t) \in N \times N} minDel_{st}} \qquad (12)$$

$$f(w) = \alpha\Phi^*(w) + (1 - \alpha)\gamma^*(w) \qquad (13)$$

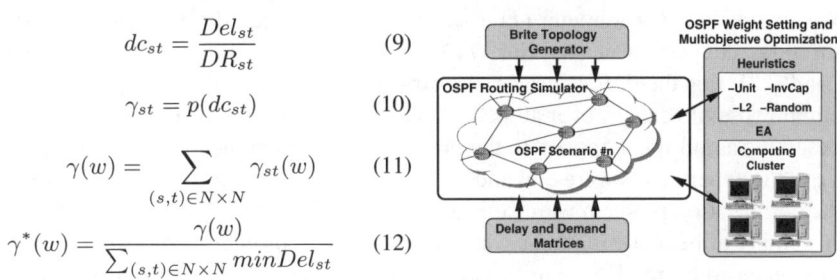

Fig. 3. Platform for performance evaluation

In this work, Evolutionary Algorithms (EAs) are proposed to address the OSPF weight setting problem, both by considering the multiobjective formulation, or by taking each of the two distinct aims separately. In the proposed EA, each individual encodes a solution as a vector of integer values, where each value (gene) corresponds to the weight of an arc in the network (values range from 1 to w_{max}). The individuals in the initial population are randomly generated, with the arc weights taken from a uniform distribution in the allowed range. In order to create new solutions, several reproduction operators were used, more specifically two mutation and two crossover operators: **Random Mutation** - replaces a given gene by a new randomly generated value, within the allowed range $[1, w_{max}]$; **Incremental/decremental Mutation** - replaces a given gene by the next or by the previous value (with equal probabilities) and constrained to respect the range of allowed values; **Uniform crossover** and **Two-point crossover** - two standard crossover operators, applied in the traditional way [12]. All operators have equal probabilities in generating new solutions. When a single objective is considered the fitness of an individual (encoding weight set w) is calculated using functions $\Phi^*(w)$ for congestion and $\gamma^*(w)$ for delays. For multiobjective optimization a simple scheme was devised, with the fitness ($f(w)$) of the individual derived by Eq. (13).

3 Experiments and Results

The conceptual model of the experimental platform that was implemented and used in this work for results evaluation is presented in Fig. 3. For this purpose, a set of

---

[2] For each pair of nodes the minimum end-to-end delay, $minDel_{st}$, is calculated as the delay of the path with minimum possible overall delay.

12 networks was generated by using the Brite topology generator [13], varying the number of nodes ($N = 30, 50, 80, 100$) and the average degree of each node ($m = 2, 3, 4$). This resulted in networks ranging from 57 to 390 links (graph edges). The link bandwidth (capacity) was generated by an uniform distribution between 1 and 10 Gbits/s. The network was generated using the Barabasi-Albert model, using a heavy-tail distribution and an incremental grow type (parameters HS and LS were set to 1000 and 100, respectively). In the generated examples, the propagation delays were assumed as the major component of the end-to-end delay of the networks paths. Thus, the network queuing delays at each network node were not considered (i.e. $Del_{st,q} = 0$). For each of the twelve network instances a set of three distinct instances of D and DR were created. A parameter (D_p) was considered which determined the expected mean of the congestion in each link (u_a) (values for D_p in the experiments were 0.1, 0.2 and 0.3). For the DR matrices, the strategy was to calculate the average of the minimum possible delays, over all pairs of nodes. A parameter (DR_p) was considered, representing a multiplier applied to the previous value to get the matrix DR (values for DR_p in the experiments were 3, 4 and 5). Overall, a set of $12 \times 3 \times 3 = 108$ instances of the optimization problem were considered.

A number of heuristic methods was implemented, to provide a comparison with the results obtained by the EAs: **Unit** - sets all arc weights to 1 (one); **InvCap** - sets arc weights to a value inversely proportional to the capacity of the link; **L2** - sets arc weights to a value proportional to the physical Euclidean distance (L2 norm) of the link; **Random** - a number of randomly generated solutions are analyzed and the best is selected, where the number of solutions considered is always equal to the number of solutions evaluated by the EA in each problem. The proposed EA, the heuristics and the OSPF routing simulator were implemented by the authors using the Java programming language. The EA was run for a number of generations ranging from 1000 to 6000, a value that was incremented proportionally to the number of variables optimized by the EA. The running times varied from a few minutes in the small networks, to a few hours in the larger ones. So, in order to perform all the tests, a computing cluster with 46 dual Xeon nodes was used. The population size was kept in 100 and the w_{max} was set to 20. Since the EA and the Random heuristic are stochastic methods, R runs were executed in each case (R was set to 10 in the experiments). For a better understanding, the results are grouped into three sets according to the cost function used. The first two consider single objective cost functions, for the optimization of congestion and delays respectively. These are used mainly as a basis for the comparison with the results obtained with the last group, that presents the results using the multiobjective cost function. In all figures presented in this section the data was plotted in a logarithmic scale, given the exponential nature of the penalty function adopted. Since the number of performed experiments is quite high, it was decided in the following sections to present all the results for just one of the networks (out of the 12) in order to explain the experimental methodology, and then to show some aggregate results to draw conclusions.

3.1   Congestion

Table 1 shows the results for the optimization of the congestion, for one of the networks (with 100 nodes and 197 links). In this table, the first column represents the demand generation parameter D_p (higher values for this parameter indicate higher mean demands, thus harder optimization problems). The remaining columns indicate the congestion measure ($\Phi^*(w)$) for the best solution (w) obtained by each of the methods considered in this study. In the case of the EAs and Random heuristic the values represent the mean value of the results obtained in the set of runs. Table 2 shows the results for all the 12 available networks, averaged by the demands levels (value of D_p), including in the last line the overall mean value for all problem instances. It is clear that the results for all the methods get worse with the increase of D_p, as would be expected. The comparison between the methods shows an impressive superiority of the EA when compared to the heuristic methods. In fact, the EA achieves solutions which manage a very reasonable behavior in all scenarios (worse case is 1.49), while the other heuristics manage very poorly. Even $InvCap$, an heuristic quite used in practice, gets poor results when D_p is 0.2 or 0.3 (Fig. 4)[3], which means that the optimization with the EAs assures good network behavior in scenarios where demands are at least 200% larger than the ones where $InvCap$ would assure similar levels of congestion.

Table 1. Optimization of congestion (Φ^*) in one network with 100 nodes and 197 links

D_p	Unit	L2	InvCap	Random	EA
0.1	3.62	190.67	1.68	12.05	1.02
0.2	136.75	658.66	135.07	280.27	1.25
0.3	264.02	874.89	488.53	551.65	1.49

Table 2. Results for the optimization of congestion (Φ^*) - averaged results by demand levels

D_p	Unit	L2	InvCap	Random	EA
0.1	8.03	215.94	1.50	75.75	1.02
0.2	99.96	771.87	57.70	498.74	1.18
0.3	227.30	1288.56	326.33	892.87	1.73
Overall	111.76	758.79	128.51	489.12	1.31

Figure 5, on the other hand, represents the results for congestion, but aggregated by the number of arcs (links). It is clear in both cases that the results obtained by the EAs are quite scalable, since the quality levels are not affected by the number of nodes or edges in the network graph. The results obtained in this section show that the EA makes an effective method for the optimization of OSPF weights, in order to minimize the congestion of the network. These results confirm the findings of other single objective OSPF optimization works (e.g. Ericsson et al [14]), although a precise comparison of the approaches is impossible since the data is not available.

3.2 Delays

Regarding the optimization of delays (cost function γ^*), a similar methodology was adopted. Indeed, in Table 3 the results for the same example network are shown. The methods used in the optimization are the same as in the previous section. In this case, the first column represents the parameter used for the generation of delay requirements (DR_p). On the other hand, Table 4 and Fig. 6 represent the results obtained for the delay optimization averaged by the parameter used in the generation of delays requirements (DR_p). In this case, the results of all methods improve when the value is higher, since higher delay requirements are easier to comply.

---

[3] In the figures the white area represents the acceptable working region whereas an increasing level of gray is used to identify working regions with increasing levels of service degradation.

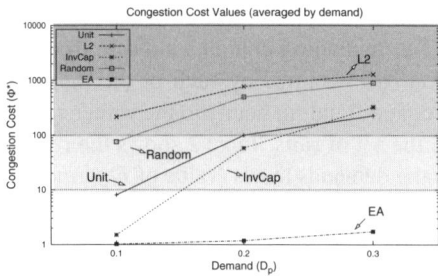

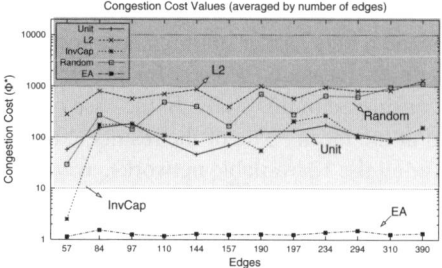

Fig. 4. Results obtained by the different methods in congestion optimization (averaged by D_p)

Fig. 5. Results obtained by the different methods in congestion optimization (averaged by edges)

Table 3. Optimization of delays (γ^*) in one example network with 100 nodes and 197 links

DR_p	Unit	L2	InvCap	Random	EA
3	13.50	1.38	201.62	4.36	1.38
4	2.00	1.13	18.33	1.82	1.13
5	1.47	1.04	3.62	1.54	1.04

Table 4. Optimization of delays (γ^*)- averaged results by the delay requirements

DR_p	Unit	L2	InvCap	Random	EA
3	152.37	2.94	577.94	156.62	2.85
4	28.78	1.25	158.85	24.35	1.25
5	6.59	1.10	44.13	4.29	1.10
Overall	62.58	1.76	260.30	61.75	1.73

The relative performance of each method shows a good behavior of the EA, as before, but now there is a simpler heuristic method - the L2 - that achieves very similar results. This is not a surprise, since in the proposed model only propagation delays were considered and these are proportional to the length of each link. The L2 heuristic considers the OSPF weights to be proportional to the arc length, which means they are also directly proportional to the delays. So, it is clear that the L2 heuristic exhibits a near-optimal behavior in this problem. It is important to notice that in the context of network management, the minimization of propagation delays, disregarding congestion, is typically not an optimization aim by itself. So, the results in this section will be used mainly as a basis for comparison with the results of multiobjective optimization. As before, the results for the delay minimization are also shown aggregated by the number of links (Fig. 7). The scalability of both L2 and the EAs prevails in these results.

3.3   Multiobjective Optimization

From the set of methods discussed before, only the EA and the Random heuristic can be used to perform multiobjective optimization by considering the optimization of function f (Equation 13) as the aim. In all other heuristic methods, the solution is built disregarding the cost function, so the results for multiobjective optimization can be pasted from the ones obtained in the previous sub-sections. The results of both EAs and Random methods are presented in terms of the values for the two objective functions (Φ^* and γ^*), since the value of f for these solutions can be easily obtained and is not relevant to the analysis (it does not represent any real measure for the network behavior). Three distinct values for α will be tested: 0.25, 0.5 and 0.75. The value of 0.5 considers each aim to be of equal importance, while the 0.25 favours the minimization of delays

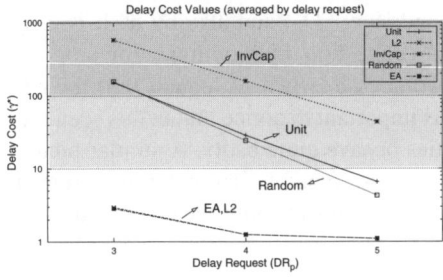
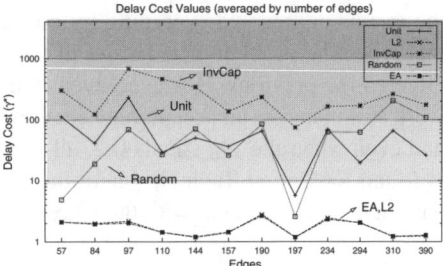

Fig. 6. Results obtained by the different methods in delay optimization (averaged by DR_p)

Fig. 7. Results obtained by the different methods in delay optimization (by the number of links)

and 0.75 will give more weight to congestion. Table 5 represents the results obtained in the example network, for the the multiobjective optimization obtained by the EAs and Random heuristics. In this table only the results for $\alpha = 0.5$ are shown. The methodology for the other values of α is essentially the same. In the table, the first two columns represent the parameters for demand and delay requirements; the next two indicate the results for the Random heuristic in both aims and, finally, the last two give the results of the EA for both congestion and delay, each with an extra information indicating the percentage by which this results exceed the ones obtained by the corresponding EA under with a single objective cost function.

Table 5. Multiobjective optimization ($\alpha = 0.5$) in one example network with 100 nodes and 197 links

D_p	DR_p	Random		EA	
		Φ^*	γ^*	Φ^* (%)	γ^* (%)
0.1	3	27.36	39.97	1.14 (11.4%)	1.52 (10.2%)
0.1	4	7.22	16.06	1.09 (6.9%)	1.26 (11.8%)
0.1	5	8.82	2.28	1.08 (6.1%)	1.13 (8.9%)
0.2	3	356.25	29.42	1.47 (17.4%)	1.75 (26.2%)
0.2	4	274.06	2.37	1.40 (11.9%)	1.42 (25.9%)
0.2	5	339.06	1.96	1.38 (9.8%)	1.29 (23.7%)
0.3	3	587.51	48.72	1.76 (18.4%)	2.04 (47.8%)
0.3	4	495.32	7.08	1.61 (8.2%)	1.56 (38.4%)
0.3	5	601.00	2.34	1.56 (5.0%)	1.37 (31.3%)

Table 6. Overall results for the multiobjective optimization - averaged by α

α	Random		EA	
	Φ^*	γ^*	Φ^* (%)	γ^* (%)
0.25	544.47	107.99	2.02 (47.2%)	2.33 (32.5%)
0.5	506.45	130.81	1.68 (25.7%)	2.49 (43.8%)
0.75	468.04	175.82	1.61 (19.5%)	2.92 (69.5%)

Table 7. Results for the multiobjective optimization - averaged by nodes

Node	Random		EA	
	Φ^*	γ^*	Φ^* (%)	γ^* (%)
30	283.32	74.77	1.58 (19.9%)	2.25 (24.3%)
50	442.16	165.63	1.78 (36.0%)	2.96 (51.9%)
80	619.14	170.75	1.62 (22.8%)	2.37 (42.7%)
100	681.17	112.09	1.75 (24.3%)	2.38 (56.2%)

In Table 6 the results obtained were aggregated averaging by the parameter α. The results shown in this table make clear its effect, once it is possible to observe different trade-offs between the two objectives. Indeed, when α increases the results on congestion improve, while the reverse happens to the delay minimization. The intermediate value of α (0.5) provides a good compromise between the two objectives. In this case, the overall results show that, in average, there is a 25% decrease in the congestion performance and around 44% in the delays minimization, both when comparing to single objective optimization. These values are quite good and, in conjunction with the average values for both cost functions, indicate an acceptable performance. Table 8 shows the results aggregated averaging by the demand level (D_p) and also by α. It is clear that when the problem gets harder in terms of congestion, both optimization aims are affected the previous. However, even in the worst case (when D_p equals 0.3) the network still manages an acceptable behavior. It is important to notice that in this scenario, and even when the D_p equals 0.2, all heuristics behave quite badly. A similar picture is found looking at Table 9, where the results are averaged by the delay requirement parameter DR_p. In fact, with the increase of DR_p the results improve on both aims.

Table 8. Multiobjective results - aver. by D_p

α	D	Random		EA	
		Φ^*	γ^*	Φ^* (%)	γ^* (%)
0.25	0.1	110.51	89.76	1.28 (25.4%)	1.85 (7.0%)
	0.2	544,74	107.08	1.64 (39.9%)	2.19 (24.6%)
	0.3	978.16	127.11	3.15 (76.3%)	2.95 (66.0%)
0.5	0.1	88.00	106.79	1.17 (14.5%)	1.92 (12.8%)
	0.2	481.50	136.68	1.47 (25.1%)	2.32 (35.2%)
	0.3	949.85	148.96	2.41 (37.5%)	3.23 (83.3%)
0.75	0.1	73.92	142.41	1.10 (8.3%)	2.12 (25.1%)
	0.2	469.42	180.58	1.35 (14.9%)	2.58 (53.6%)
	0.3	914.79	204.49	2.38 (35.1%)	4.05 (129.9%)

Table 9. Multiobjective results - aver. by DR_p

α	DR	Random		EA	
		Φ^*	γ^*	Φ^* (%)	γ^* (%)
0.25	3	616.36	246.07	2.63 (80.8%)	4.03 (46.2%)
	4	536.07	60.37	1.77 (34.2%)	1.60 (28.0%)
	5	480.98	17.52	1.67 (26.5%)	1.36 (23.4%)
0.5	3	535.28	283.16	1.95 (42.8%)	4.22 (55.2%)
	4	505.69	82.04	1.59 (20.3%)	1.78 (41.8%)
	5	478.37	27.23	1.51 (14.2%)	1.48 (34.4%)
0.75	3	506.84	372.04	1.89 (37.1%)	5.05 (94.1%)
	4	468.14	116.96	1.48 (11.6%)	2.03 (62.3%)
	5	483.14	38.48	1.46 (9.7%)	1.68 (52.2%)

Table 7, on the other hand, confirms the good scalability properties of the EA. In fact, and as seen in the previous sections for both congestion and delay optimization, the results are almost constant for the different network sizes (in this case, measured by the number of nodes). A different view is offered by Figs. 8 and 9 where the results are plotted with the two objective functions in each axis. The former shows the results averaged by the demand levels and the latter by the delay requirements parameter. In both cases the value considered for α was 0.5, although the overall view for different values of this parameter would be very similar. In these graphs, the good overall network behavior of the solutions provided by the EA is clearly visible, both in absolute terms, regarding the network behavior in terms of congestion and delays, and when compared to all other alternative methods. In fact, it is easy to see that no single heuristic is capable of acceptable results in both aims simultaneously. L2 behaves well in the delay minimization but fails completely in congestion; InvCap is better on congestion (although in a very limited range) but fails completely in the delays. EAs, on the other hand, are capable of a good compromise between both optimization targets.

Finally, Figs. 10 and 11 show similar graphs but considering only the EAs and plotting the results for different values of α. In the figures the trade-offs between the two objectives are clear. Regarding to Fig. 10 the obtained delay and congestion cost values are averaged for distinct values of traffic demands ($D_p = 0.1, 0.2$ and 0.3). Moreover, three distinct lines are plotted, each one representing the results obtained assuming distinct values of α (0.25, 0.5 and 0.75). As observed in Fig. 10, the results show the correctness of the proposed optimization model. In fact, within a given demand value, the plots are shifted towards the upper left region of the graph as the α value increases. This behaviour corroborates the optimization model underpinning concept, in which higher values of α lead to an improvement in the congestion metric but, at the same time, a penalization in the delay performance. The results plotted in Fig. 11 show the

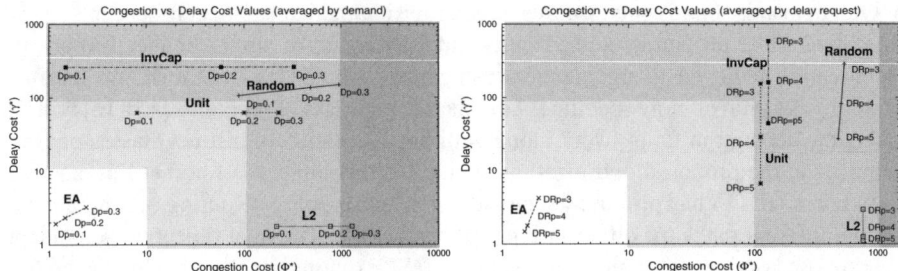

Fig. 8. Results of the different methods in the multiobjective optimization (for $\alpha = 0.5$)

Fig. 9. Results of different methods in the multiobjective optimization (for $\alpha = 0.5$)

obtained delay and congestion cost values averaged now for distinct values of the delay requests ($DR_p = 3, 4$ and 5). As in the case of Fig. 10, the results clearly show the correctness of the system dynamics, as the delay and congestion performance of distinct experimental scenarios is controlled by the α parameter. As observed, network configuration assuming lower values for α achieve a better delay performance.

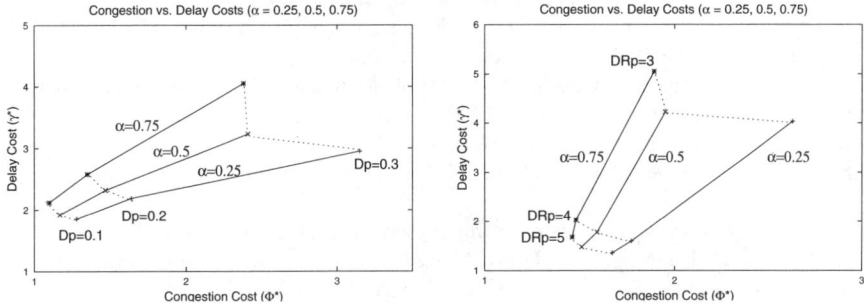

Fig. 10. Results of the EAs (averaged by D_p)   **Fig. 11.** Results of the EAs (averaged by DR_p)

4   Conclusions and Further Work

This work presented an optimization scheme based on Evolutionary Algorithms with an integer representation for the purpose of multiobjective routing in the Internet. To achieve this aim, an analytical model was developed allowing the performance evaluation of several QoS constrained OSPF routing scenarios of a given ISP. Resorting to a large set of network topology configurations, each one constrained by several bandwidth and end-to-end delay requirements, it was shown that the proposed EAs were able to provide OSPF weight settings able to satisfy the users demands and outperform common weight setting heuristics.

The research results presented in this work give ground to the idea that it is possible to develop network management tools which automatically provide network administrators with optimal configurations for a given network topology and corresponding sets of QoS demands. In this way, ISP resource provisioning management tasks can be now simplified while providing better results and, consequently, strong financial improvements can be achieved by organizations using the proposed OSPF optimization scheme. The consideration of more specific EAs to handle this class of problems [15][16] will be taken into account in future work along with the integration of distinct classes of QoS demands in the proposed optimization model. On this topic, the Internet Engineering Task Force (IETF) has proposed standards on Multi-topology Routing aiming at providing different paths for different types of traffic [17]. The final objective is to adapt the proposed optimization model to deal with OSPF routing schemes having the ability to use multiple levels of weights.

References

1. Zheng Wang. *Internet QoS: Architectures and Mechanisms for Quality of Service*. Morgan Kaufmann Publishers, 2001.
2. D. Verma. *Supporting Service Level Agreement on IP Networks*. McMillan Publishing, 1999.
3. B. Davie and Y. Rekhter. *MPLS: Multiprotocol Label Switching Technology and Applications*. Morgan Kaufmann, USA, 2000.
4. J. Moy. RFC 2328: OSPF version 2, April 1998.
5. T.M. ThomasII. *OSPF Network Design Solutions*. Cisco Press, 1998.
6. B. Fortz and M. Thorup. Internet Traffic Engineering by Optimizing OSPF Weights. In *Proceedings of IEEE INFOCOM*, pages 519–528, 2000.
7. A. Medina et al. Traffic matriz estimation: Existing techniques and new directions. *Computer Communication Review*, 32(4):161–176, 2002.
8. Ravindra et al. *Network Flows*. Prentice Hall, 1993.
9. E. W. Dijkstra. A note on two problems in connexion with graphs. *Numerische Mathematik*, 1(269-271), 1959.
10. J. Moy. *OSPF, Anatomy of an Internet Routing Protocol*. Addison Wesley, 1998.
11. G. Bolch et al. *Queueing Networks and Markov Chains - Modeling and Performance Evaluation with Computer Science Applications*. Jhon Wiley and Sons INC., 1998.
12. Z. Michalewicz. *Genetic Algorithms + Data Structures = Evolution Programs*. Springer-Verlag, USA, third edition, 1996.
13. A. Medina et al. BRITE: Universal Topology Generation from a User's Perspective. Technical Report 2001-003, January 2001.
14. M. Ericsson, M.G.C. Resende, and P.M. Pardalos. A Genetic Algorithm for the Weight Setting Problem in OSPF Routing. *J. of Combinatorial Optimization*, 6:299–333, 2002.
15. C.M. Fonseca and P.J. Fleming. An Overview of Evolutionary Algorithms in Multiobjective Optimization. *Evolutionary Computation*, 3(1):1–16, 1995.
16. C.A. Coello Coello. *Recent Trends in Evolutionary Multiobjective Optimization*, pages 7–32. Springer-Verlag, London, 2005.
17. P. Psenak et al. Multi-topology (mt) routing in ospf (internet draft), January 2006.

Probabilistic QoS Guarantees with FP/EDF Scheduling and Packet Discard in a Real Time Context: A Comparative Study of Local Deadline Assignment Techniques

Fadhel Karim Maïna and Leila Azouz Saïdane

Ensi, laboratoire Cristal, University of Manouba, Tunisia
maina_fadhel@voila.fr, leila.saidane@ensi.rnu.tn

Abstract. In this paper, we are interested in comparing local deadline assignment techniques in a multi-hop network supporting real time traffic with end-to-end delay constraints, when the FP/EDF scheduling is used, assuming that packets which don't respect their local delay constraints are discarded. In each node, packets are scheduled according to their Fixed Priorities (FP), and within the same priority, packets are scheduled according to the Earliest Deadline First (EDF) policy, using local deadlines, which correspond to the sojourn times not to be exceeded in that node. Consequently, an accurate choice of these local deadlines must be done in order to respect the flows' end-to-end delay constraints and minimize the packet discard rate. As we are interested in giving probabilistic QoS guarantees, we develop a mathematical model to compare the performances of five existing deadline assignment techniques. We show that all these techniques give very high packet discard rates. So, we propose to use another packet discard policy and we show that it gives better results.

1 Introduction

Fixed Priority (FP) scheduling has been extensively studied in the last years [1, 2]. It is well adapted for service differentiation and above all, it is easy to implement. However, with such a scheduling, the number of available priorities is usually lower than the number of flows in the network. Consequently, several flows belonging to different types of applications will share a same priority. If, within a same priority, all flows are served in te same way, real-time flows (VoIP, videoconference, etc.), which are characterized by stringent end-to-end delay constraints, may be penalized. That's why we propose to introduce a Dynamic Priority (DP) scheduling within a same priority. Such a scheduling policy is called FP/DP. In our case, we propose to use the Earliest Deadline First (EDF) as a DP scheduling, since it is known to provide the optimal delay performance in the uniprocessor case [3]. EDF also gives better statistical QoS guarantees than other scheduling algorithms such as GPS (General Processor Sharing) as shown in [4].

G. Parr, D. Malone, and M. Ó Foghlú (Eds.): IPOM 2006, LNCS 4268, pp. 49–60, 2006.

According to EDF, flow packets are scheduled in a node according to their "absolute deadlines". The absolute deadline of a packet is the sum of its arrival instant to the node, plus a "relative deadline", also called "local deadline", which corresponds to the maximum sojourn time of the packet in that node. So, an accurate choice of the packets' local deadlines must be done in order to meet their end-to-end delay constraints, also called end-to-end deadlines. We address here the problem of deadline assignment.

We are also interested in giving flows a probabilistic QoS guarantee, expressed in terms of end-to-end deadline miss probability. In fact, it is known that a deterministic quantitative QoS guarantee is generally pessimistic, and yields to an over-provisioning of network resources. This statement remains true for the deterministic study of FP/EDF developed in [5]. Frameworks for probabilistic delay guarantees can operate in the setting of either stochastic (i.e markovian) or regulated (i.e leaky bucket) source models. Numerous frameworks that consider stochastic (see [6] for EDF and [2] for PQ) or regulated (see [7] and [4] for EDF and GPS) source models have been developed in literature. All these frameworks allow to compute the deadline miss probability only for a flow aggregate. In this paper, we develop a stochastic source model framework which also allows the computation of the deadline miss probability for single flows as well as for a class of flows.

We also assume in this study that packets which miss their local deadlines at any node are discarded. We show that in this case, the end-to-end deadline miss probability of a given flow corresponds to the probability that packets belonging to this flow are discarded at each crossed node. Consequently, the probabilistic QoS guarantee can be given in terms of packet discard rate.

We then compare the performances of five existing deadline assignment techniques ("pure laxity", "normalized laxity", "fair Assignment", "Proportional to the bandwidth" and "Proportional to the workload") in terms of packet discard rates. We show that all these techniques discard a high rate of packets, which wouldn't fit the real-time applications that are sensitive to packet losses. So, we propose to use another packet discard policy to minimize the packet discard rate while maintaining, for each flow, the respect of its end-to-end delay constraint.

This paper is organized as follows: in section 2, we present the adopted assumptions. In section 3, we focus on the deadline assignment problem and the five deadline assignment techniques adopted for our study. In section 4, we present our mathematical model. In section 5, we show how to compute the end-to-end deadline miss probability for a flow. And, in section 6, we compare the performances of the five adopted deadline assignment techniques and show the advantages of the new packet discard policy.

2   General Assumptions

2.1   Scheduling Model

We consider that all nodes in the network schedule packets according to the non-preemptive (FP/EDF) scheduling. Packets are first scheduled according to their

Fixed Priorities (FP), and within a same priority, they are scheduled according to the (EDF). With (EDF), packets are scheduled according to their "absolute deadlines". The absolute deadline of a packet p_k from a flow τ_k in a node i is the sum of its arrival instant t_k to this node, plus a "relative deadline" E_k^i obtained from the end-to-end deadline E_k of the flow by a deadline assignment method. We then adopt the following assumption:

Assumption 1: Packets whose absolute deadlines have expired are discarded at the node's scheduler.

2.2 Network Model

We adopt the following assumptions concerning the considered network.

Assumption 2: Links interconnecting nodes are supposed to be $FIFO$.

Assumption 3: The transmission delay between two nodes is supposed to be negligible.

Assumption 4: Network is reliable, that is, neither network failures nor packet losses are considered.

2.3 Traffic Model

We consider a set $S = (\tau_1, \tau_2, ..., \tau_K)$ of K flows sharing R priorities. Each flow is characterized by its fixed priority and its end-to-end deadline. In addition, we suppose that:

Assumption 5: Each flow is assumed to follow a fixed sequence of nodes, called path.

For instance, $MPLS$ can be used to fix the path followed by a flow.

3   Deadline Assignment Methods

3.1   The Deadline Assignment Problem Formalization

Consider a network characterized by a directed graph $G(Q, L)$, where Q denotes a set of nodes and L a set of links. This network is crossed by a set $\mathbb{S} = (\tau_1, \tau_2, ..., \tau_K)$ of flows. Let q_1 and q_n be any two nodes of Q, let $\mathcal{P}(q_1, q_n)$ be a path from node q_1 to node q_n composed of $\{l_1, l_2, ...l_{n-1}\}$ links. $\mathcal{P}(q_1, q_n)$ can be followed by more than one flow. Let E_k be a positive constant denoting the end-to-end deadline of a flow $\tau_k \in \mathbb{S}$ following $\mathcal{P}(q_1, q_n)$. Each link $l_i \in \mathcal{P}(q_1, q_n)$ can provide to the τ_k a QoS level chracterized by two metrics: a local deadline E_k^i and a probability P_k^i of missing this deadline. The problem consists in finding for each link $l_i \in \mathcal{P}(q_1, q_n)$ the QoS level meeting $\sum_{i=1}^{n-1} E_k^i \leq E_k$ and minimizing $1 - \prod_{i=1}^{n-1}(1 - P_k^i)$. A deterministic version of this problem can be found in [8].

3.2   Deadline Assignment Methods

[9] and [10] propose two methods to assign the local deadline E_k^i , of a flow $\tau_k \in S$ in the link l_i, from its end-to-end deadline E_k. In both methods, the local deadline is equal to T_k^i, the service duration of the flow τ_k in the link l_i, plus a laxity obtained by a metric. Two laxity metrics are used: *pure laxity* R_{pure} and *normalized laxity* R_{norm}.

- **pure laxity** provides a fair laxity assignment where N denotes the number of links on the path of flow τ_k, with $R_{pure} = \dfrac{E_k - \sum\limits_{j=1}^{N} T_k^j}{N}$ and $E_k^i = T_k^i + R_{pure}$
- **normalized laxity** provides a laxity assignment proportional to the local execution time: $R_{norm} = \dfrac{E_k - \sum\limits_{j=1}^{N} T_k^j}{\sum\limits_{j=1}^{N} T_k^j}$ and $E_k^i = T_k^i(1 + R_{norm}) = \dfrac{T_k^i}{\sum\limits_{j=1}^{N} T_k^j} E_k$.

[11] proposes three deadline assignment methods in the case of a network composed of N heterogeneous segments:

- The first method is the **fair deadline assignment**. With this method, the local deadline assigned to flow τ_k on segment l_i is $E_k^i = E_k/N$. Note that when the execution duration of flow τ_k is identical on all nodes, R_{pure} and R_{norm}. also give the same value of the local deadline E_k^i.
- The second method assigns the local deadlines **proportionally to the segment workload**. With this method, the local deadline assigned to flow τ_k on segment l_i is $E_k^i = \dfrac{U_i}{\sum\limits_{j=1}^{N} U_j} E_k$ where U_i denotes the workload of segment l_i.
- The third method assigns the local deadline **inversely proportional to the available bandwidth** on segment l_i, leading to $E_k^i = \dfrac{\sum\limits_{j=1}^{N} B_j - B_i}{(N-1)\sum\limits_{j=1}^{N} B_j} E_k$ where B_i is the available bandwidth on segment l_i.

In our study, we consider these five deadline assignment methods and we compare their performances in terms of packet discard rates.

4   Mathematical Model

We focus on K flows sharing R priorities in a network composed of a set of Q edge and core routers. Let R be the highest priority. It is known that the arrivals of flow packets aren't a Poisson process. However, [12] has shown that considering a Poisson process at flow packet scale is an upper bound of the reality. We can then introduce this assumption to simplify the study.

Assumption 7: External flow packet arrivals (arrivals to the network) are supposed to be independent Poisson processes.

We denote by λ_{rk} the Poisson arrival rate to the network of a class k flow with priority r. Each link l_{ab} connecting node a to node b, where $a, b \in Q$ is characterized by its bandwidth V_{ab}. Each flow τ_k, having the priority r is characterized by its packet processing time, its packet inter arrival rate, its fixed priority and its end-to-end deadline. Packets belonging to the same flow, inherit from its characteristics. According to assumption 5, all the packets of a same flow follow the same path. Flow paths are defined by the matrix $F = (F_{kq})$ $k=1...K, q=1...Q$ where F_{kq} is the row of the node q in the path of flow τ_k. If τ_k doesn't cross the node q, then $F_{kq} = 0$. Hence we define δ_{rk}^{ab} by:

$$\delta_{rk}^{ab} = \begin{cases} 1 & \text{if } F_{kb} - F_{ka} = 1 \quad \text{with } F_{kb} > 0, F_{ka} > 0 \text{ and } V_{ab} > 0 \\ 0 & \text{else} \end{cases} \tag{1}$$

that is $\delta_{rk}^{ab} = 1$, if the link l_{ab} belongs to the path of the flow τ_k having the priority r. Each node can be considered as a set of queueing systems. Arriving packets are stocked in a first file to be processed and switched over the appropriate link. Each link corresponds to a queueing system, where the service is the transmission of a packet over this link. By supposing that the processing time at the first file is instantaneous, the node response time, for a packet going through node a to node b, corresponds to the response time of the queueing system modelling the link l_{ab}. Let $\widetilde{s_{rk}^{ab}}$ be the random variable associated to the sojourn time of class k packets with priority r at the link l_{ab}. According to assumption 1, packets belonging to the flow τ_k with the priority r are discarded with the probability $P[\widetilde{s_{rk}^{ab}} > E_k^{ab}] = 1 - P[\widetilde{s_{rk}^{ab}} \leq E_k^{ab}]$. To simplify this study, we introduce the following assumptions:

Assumption 8: The sojourn times of the packets inside the network are supposed to be stochastically independent.

Bonald et al. assume in [12] that this is a worst case assumption which only leads to upper bounds on performance.

Assumption 9: We then suppose that packet arrivals of a flow τ_k with priority r, to a link l_{ab} are also a Poisson process with the parameter

$$\lambda_{rk}^{ab} = \delta_{rk}^{ab} \lambda_{rk} \left(\prod_{\substack{i, j \text{ such that } F_{kb} > F_{kj}, \\ F_{kj} > F_{ki} \text{ and } \delta_{rk}^{ij} = 1}} \left(1 - P[\widetilde{s_{rk}^{ij}} > E_k^{ij}] \right) \right) \tag{2}$$

where δ_{rk}^{ab} is given by 1. λ_{rk}^{ab} is the part of λ_{rk} which was not discarded at all the links l_{ij} preceeding the link l_{ab}.

According to the traffic description, to assumption 9 and to the FP/EDF scheduling, each link l_{ab} can be modelled by an $M/G/1$ station with K classes of customers (the K flows), and the non-preemptive FP/EDF (with R priorities) discipline. Let ρ_{rk}^{ab} be the utilization factor of the link l_{ab} server by the packets of class k with priority r. ρ_{rk}^{ab} is given by

$$\rho_{rk}^{ab} = \lambda_{rk}^{ab} \overline{X_{rk}^{ab}} \tag{3}$$

where $\overline{X_{rk}^{ab}}$ corresponds to the average service time of packets of a flow τ_k with priority r at the link l_{ab} and is given by:

$$\overline{X_{rk}^{ab}} = \frac{\beta_{rk}}{V_{ab}} \tag{4}$$

where β_{rk} is the average packet size of such a flow and V_{ab} is the bandwidth of the link l_{ab}. The utilization factor ρ^{ab} of the link l_{ab} server is then given by:

$$\rho^{ab} = \sum_{r=1}^{R}\sum_{k=1}^{K}\rho_{rk}^{ab} \tag{5}$$

5   Deadline Miss Probability Computation

5.1   Local Deadline Miss Probability Computation

To compute for any flow τ_k with priority r the probability $P[\widetilde{s_{rk}^{ab}} > E_k^{ab}]$ of missing its local deadline E_k^{ab} at the link l_{ab}, we must evaluate the distribution function of $\widetilde{s_{rk}^{ab}}$. This distribution function is obtained by inspecting its Laplace transform, denoted $\left(S_{rk}^{ab}\right)^*(s)$, and given by

$$\left(S_{rk}^{ab}\right)^*(s) = \left(W_{rk}^{ab}\right)^*(s) \times \left(B_{rk}^{ab}\right)^*(s) \tag{6}$$

where $\left(B_{rk}^{ab}\right)^*(s)$ is the Laplace transform (LT) of the service time probability density function of packets from class k with priority r at the link l_{ab} and $\left(W_{rk}^{ab}\right)^*(s)$ is the LT of the waiting time density of packets from class k with priority r at the link l_{ab}.

We focus on the computation of $\left(W_{rk}^{ab}\right)^*(s)$. A packet m from class k with priority r and a relative deadline E_k^{ab} has to wait for:

1. The packet found in service,
2. packets belonging to flows with priorities greater than r and which arrive to the link l_{ab} before m starts its execution,
3. and packets having the priority r with an absolute deadline D, such that $D \leq t_k + E_k^{ab}$, which arrive to the link l_{ab} before m starts its processing.

To compute $\left(W_{rk}^{ab}\right)^*(s)$, we define two classes of packets, as in [13]:

- $\mathfrak{C}_{rk}^+$ which represents the packets found in the system upon the arrival of m with priorities greater than or equal to r, and having to be served before m (called priority packets)
- and $\mathfrak{C}_{rk}^-$ which represents the packets found in the system upon the arrival of m with priority smaller than or equal to r, to be served after m (called ordinary packets).

The Poisson arrival rate of the class $\mathfrak{C}_{rk}^{+}$ packets is then given by:

$$\lambda_{rk}^{+} = \sum_{i=r+1}^{R} \sum_{j=1}^{K} \left(\lambda_{ij}^{ab}\right) + \sum_{i=1}^{K} P(i \text{ prec. } k \ / \ i \text{ arr. bef. } k) \left(\lambda_{ri}^{ab}\right) \tag{7}$$

where $P(i \text{ prec. } k \ / \ i \text{ arr. bef. } k)$ is the probability that a packet of class i and priority r, with a relative deadline E_i^{ab}, is served before a packet from flow k (that is the probability that $t_i + E_i^{ab} < t_k + E_k^{ab}$ knowing that $t_i < t_k$). According to assumption 8, the arrival times of class i packets are seen, by class k packets as they were uniformly distributed. So, we have:

$$P(i \text{ prec. } k \ / \ i \text{ arr.bef. } k) = \begin{cases} 1 \text{ if } E_i^{ab} \leq E_k^{ab} \\ \int\limits_{0}^{+\infty} \max(0, \frac{t - \delta_{ik}^{ab}}{t}) \ \lambda_{rk}^{ab} e^{-\lambda_{rk}^{ab}t} dt \text{ else} \end{cases} \tag{8}$$

On the other hand, the Poisson arrival rate of the class $\mathfrak{C}_{rk}^{-}$ packets is given by:

$$\lambda_{rk}^{-} = \sum_{i=1}^{r-1} \sum_{j=1}^{K} \left(\lambda_{ij}^{ab}\right) + \sum_{i=1}^{K} [1 - P(i \text{ prec. } k \ / \ i \text{ arr.bef. } k)] \ \left(\lambda_{ri}^{ab}\right) \tag{9}$$

The LTs of the service time distribution of priority packets $(\mathfrak{C}_{rk}^{+})$ and ordinary packets $(\mathfrak{C}_{rk}^{-})$ are respectively

$$(B_{rk}^{+})^{*}(s) = \frac{1}{\left(\lambda_{rk}^{+}\right)} \left[\sum_{i=r+1}^{R} \sum_{j=1}^{K} \left(\lambda_{ij}^{ab}\right) \ (B_{ij}^{ab})^{*}(s) + \right. \tag{10}$$

$$\left. \sum_{i=1}^{K} P(i \text{ prec. } k \ / \ i \text{ arr.bef. } k) \left(\lambda_{ri}^{ab}\right) (B_{ri}^{ab})^{*}(s) \right]$$

and

$$(B_{rk}^{-})^{*}(s) = \frac{1}{\left(\lambda_{rk}^{-}\right)} \left[\sum_{i=1}^{r-1} \sum_{j=1}^{K} \left(\lambda_{ij}^{ab}\right) \ (B_{ij}^{ab})^{*}(s) + \right.$$

$$\left. \sum_{i=1}^{K} [1 - P(i \text{ prec. } k \ / \ i \text{ arr.bef. } k)] \ \left(\lambda_{ri}^{ab}\right) (B_{ri}^{ab})^{*}(s) \right] \tag{11}$$

Let $\left(W_{rk}^{+}\right)^{*}(s)$ be the LT of the waiting time distribution of priority packets (packets belonging to $\mathfrak{C}_{rk}^{+}$). $\left(W_{rk}^{+}\right)^{*}(s)$ is given by:

$$\left(W_{rk}^{+}\right)^{*}(s) = \frac{\left(1 - \rho^{ab}\right) s + \lambda_{rk}^{-} \left(1 - \left(B_{rk}^{-}\right)^{*}(s)\right)}{s - \lambda_{rk}^{+} + \lambda_{rk}^{+} \left(B_{rk}^{+}\right)^{*}(s)} \tag{12}$$

where ρ^{ab} is given by 5.(see [13] for more details). The waiting time at the link l_{ab}, of a packet m of class k with priority r corresponds to W_{rk}^{+} and the sum of the service times of packets that arrive during the delay busy period initiated by W_{rk}^{+} and which have to be served before m. The Poisson arrival rate of these packets, denoted by λ_{rk}^{++} is given by:

$$\lambda_{rk}^{++} = \sum_{i=r+1}^{R} \sum_{j=1}^{K} \left(\lambda_{ij}^{ab} \right) + \sum_{i=1}^{K} P(i \text{ prec. } k \;/\; i \text{ arr.aft. } k) \left(\lambda_{ri}^{ab} \right) \qquad (13)$$

where $P(i \text{ prec. } k \;/\; i \text{ arr.aft. } k)$ is the probability that a packet of flow i with a relative deadline E_i^{ab}, is served before our tagged packet of flow k (that is the probability that $t_i + E_i^{ab} < t_k + E_k^{ab}$ knowing that $t_i > t_k$). We have:

$$P(i \text{ prec. } k \;/\; i \text{ arr.aft. } k) = \begin{cases} 0 \text{ if } E_i^{ab} \geq E_k^{ab} \\ \int_0^{+\infty} \left(\frac{\delta_{ki}^{ab}}{E_k^{ab}} \right) \lambda_{rk}^{ab} e^{-\left(\lambda_{rk}^{ab}\right)t} dt = \frac{\delta_{ki}^{ab}}{E_k^{ab}} \text{ else} \end{cases} \qquad (14)$$

Hence, the LT of the waiting time density $\left(W_{rk}^{ab} \right)^{*}(s)$ is given by

$$\left(W_{rk}^{ab} \right)^{*}(s) = \left(W_{rk}^{+} \right)^{*} \left(s + \lambda_{rk}^{++} - \lambda_{rk}^{++} \left(\theta_{rk}^{++} \right)^{*}(s) \right) \qquad (15)$$

where $\left(\theta_{rk}^{++} \right)^{*}(s)$ is the LT of a busy period delay distribution generated by the packets which arrive after m and which have to be served before it. $\left(\theta_{rk}^{++} \right)^{*}(s)$ is the solution of the equation

$$\left(\theta_{rk}^{++} \right)^{*}(s) = \left(B_{rk}^{++} \right)^{*} \left(s + \lambda_{rk}^{++} - \lambda_{rk}^{++} \left(\theta_{rk}^{++} \right)^{*}(s) \right) \qquad (16)$$

where $\left(B_{rk}^{++} \right)^{*}(s)$ is the LT of the service time distribution $\left(b_{rk}^{++}(x) \right)$ of packets arriving during the delay busy period initiated by W_{rk}^{+} and which have to be served before m.

5.2 End-to-End Deadline Miss Probability Computation

According to assumption 3 and to the deadline assignment problem formalization, any packet belonging to the flow τ_k with priority r, which reaches its destination (is not discarded) respects certainly its end-to-end deadline (as $\sum_{i=1}^{n-1} E_k^i \leq E_k$). So, the end-to-end deadline miss probability P_{rk}^{miss} corresponds to the probability of not being discarded at any crossed node, and can be computed as follows:

$$P_{rk}^{miss} = 1 - \prod_{1 \leq i,j \leq Q} \delta_{rk}^{ij} (1 - P[\widetilde{s_{rk}^{ij}} > E_k^{ij}]) \qquad (17)$$

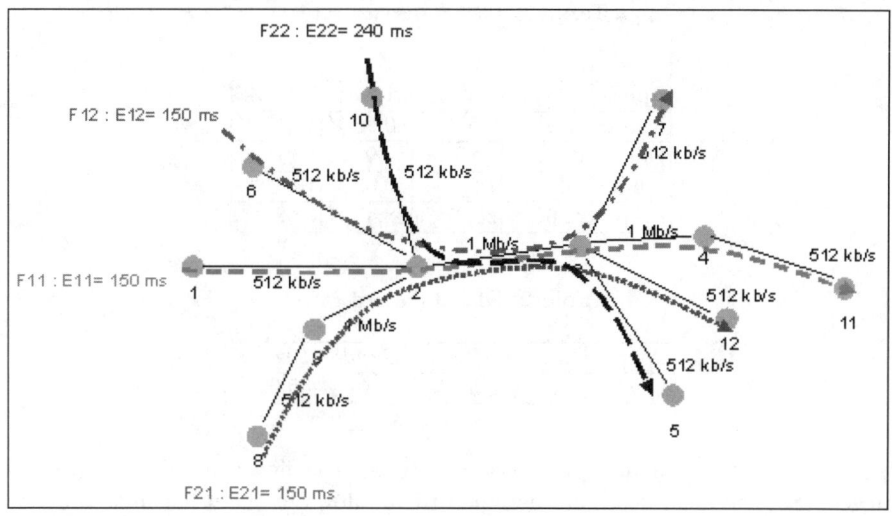

Fig. 1. Network and flows description

6   Comparison of the Different Methods

6.1   Network Description

We consider the network of Fig 1.

In this network, we consider 2 priority levels and 4 flows, τ_1, τ_2, τ_3 and τ_4 distributed over the two priority levels as follows:

- Flows τ_1 and τ_2 belong to priority p_1, and will be noted τ_{11} and τ_{12}. Their end-to-end deadlines are $E_1 = 150ms$ and $E_2 = 150ms$.
- Flows τ_3 and τ_4 belong to priority p_2, and will be noted τ_{23} and τ_{24}. Their end-to-end deadlines are $E_3 = 150ms$ and $E_4 = 240ms$.

The paths followed by these flows are given in figure 1. We study the network when the load of the link $l_{2,3}$, which is the bottleneck link, equals $0, 8$.

6.2   End-to-End Miss Probabilities

In Table 1, we present the packet discard rates (end-to-end miss probabilities) of the four flows, with the different deadline assignment techniques. We notice that the lowest packet discard rate is not given by the same technique for the four flows. In other words, there isn't a technique that suits all flows. So, to have a valuable comparison criterion, we introduce the notion of global discard rate, which corresponds to the probability that all flows miss their end-to-end deadlines. In Table 2, we present the global discard rates obtained by the five deadline assignment techniques.

It is intresting to notice that the "proportional to the workload" technique gives the highest global packet discard rate. In fact, as the load of the link

Table 1. Packet discard rates

	τ_{11}	τ_{12}	τ_{23}	τ_{24}
Rpure	9,29 %	1,69 %	31,54 %	7,49 %
Rnorm	8,09 %	2,58 %	30,96 %	17,13 %
Fair	10,57 %	1,85 %	32,94 %	6,61 %
Band	8,49 %	1,57 %	30,83 %	10,81 %
Prop	22,69 %	25,07 %	18,47 %	6,02 %

Table 2. Global discard rates

Rpure	Rnorm	Fair	Band	Prop
43,53 %	48,78 %	45,03 %	44,44 %	55,62 %

$l_{2,3}$ is high, the local deadlines assigned to all the flows in this link are the biggest. So, the local deadlines assigned to the flows in the other links are very small and the probability to miss them is very high. This leads to a high packet discard rates in these links, and consequently to a high global discard rate. In our configuration, the best results are respectively given by the "pure laxity" and the "proportional to the bandwidth" techniques. Nevertheless, we notice that all the deadline assignment techniques give very high global discard rates (more than 40%). This could be very penalizing for real-time applications which are very sensitive to packet losses. So, we propose to introduce another packet discard policy in order to minimize the packet discard rates.

6.3   The New Packet Discard Policy

We propose to discard flow packets only when their sojourn time into the network exceeds their end-to-end deadlines. Packets belonging to the flow τ_k with priority r are discarded at a link l_{ab} with the probability:

$$P[(\widetilde{s_{rk}^{ab}} + \sum_{\substack{i, j \text{ such that } F_{kb} > F_{kj}, \\ F_{kj} > F_{ki} \text{ and } \delta_{rk}^{ij} = 1}} \widetilde{s_{rk}^{ij}}) > E_k \; / \; (\sum_{\substack{i, j \text{ such that } F_{kb} > F_{kj}, \\ F_{kj} > F_{ki} \text{ and } \delta_{rk}^{ij} = 1}} \widetilde{s_{rk}^{ij}}) \leq E_k] \quad (18)$$

It is also important to say that, whith this packet discard policy, all packets which reach their destination (are not discarded) certainly respect their end-to-end deadlines.

In Table 3, we present the packet discard rates (end-to-end miss probabilities) of the four flows, with the different deadline assignment techniques, as well as the global packet discard rates obtained with the new packet discard policy.

We note that the packet discard rates are much lower than the ones obtained by the precedent discard policy. We explain this by the fact that packets which miss their local deadlines in a node recover their lateness in the next nodes. We note that the five deadline assignment techniques give close results, but the best result is given by the "proportional to the bandwidth" technique. We then

Table 3. Packet discard rates with the new policy

	τ_{11}	τ_{12}	τ_{23}	τ_{24}
Rpure	9,29 %	1,69 %	31,54 %	7,49 %
Rnorm	8,09 %	2,58 %	30,96 %	17,13 %
Fair	10,57 %	1,85 %	32,94 %	6,61 %
Band	8,49 %	1,57 %	30,83 %	10,81 %
Prop	22,69 %	25,07 %	18,47 %	6,02 %

recommend to use this technique to assign the local deadlines to flows. Similar results have been obtained with other network configurations and simulations are being done to corroborate these results.

7 Conclusion

In this paper, we focused on five deadline assignment techniques: "pure laxity", "normalized laxity", "fair assignment", "proportional to the bandwidth" and "proportional to the workload". A mathematical model was developed to compare their performances in terms of end-to-end miss probability within the FP/EDF scheduling, assuming that packets wich don't respect their local deadlines are discarded. The main contribution of this model is the possibility to compute the deadline miss probabilities for single flows or a class of flows.

In the chosen network configuration, we noticed that the five techniques give very high packet discard rates. We then proposed the use of a new discard policy where the packets are discarded only when their sojourn time into the network exceeds their end-to-end deadlines. We show that this discard policy gives better results. We also show that with this policy, the five deadline assignment techniques give close results, but the best result is given by the "proportional to the bandwidth" technique. Simulations are now being done in order to corroborate these results with a bigger network and a higher number of flows.

References

[1] K. Tindell, A. Burns, A. J. Wellings, "Analysis of hard real-time communications, Real-Time Systems", Vol. 9, pp. 147-171, 1995.

[2] E. D. Knightly, "Enforceable quality of service guarantees for bursty traffic streams", Proc. of INFOCOM'98, pp. 635-642, San Francisco, CA, March 1998.

[3] J. Liu, "Real-time systems", Prentice Hall, New Jersey, 2000.V. Sivaraman and F. Chiussi, "Statistical analysis of delay bound violations at an earliest deadline first (EDF) scheduler", Perform. Eval., vol 36-37, no. 1, pp. 457-470, 1999.

[4] V. Sivaraman, F. M. Chiussi, M. Gerla, "End-to-end statistical delay service under GPS and EDF scheduling: a comparison study", Proc. of IEEE INFOCOM'01, April 2001.

[5] S. Martin, P. Minet, L. George. "FP/EDF, a non-preemptive scheduling combining fixed priorities and deadlines: uniprocessor and distributed cases", INRIA, April 2004.

[6] V. Sivaraman and F. Chiussi, "Statistical analysis of delay bound violations at an earliest deadline first (EDF) scheduler", Perform. Eval., vol 36-37, no. 1, pp. 457-470, 1999.

[7] V. Sivaraman and F. Chiussi, "Providing end-to-end statistical delay guarantees with earliest deadline first scheduling and per-hop traffic shaping". Proc. of INFOCOM 2000, Tel Aviv, Israel, March 2000.

[8] D. Marinca and P. Minet, "Analysis of deadline assignment in distributed real-time systems", INRIA, Mach 2004.

[9] M. Di Natale, J. A. Stankovic, "Dynamic end-to-end guarantees in distributed real time systems", Proc. of Real-Time Systems Symposium, San Juan, Puerto Rico, 7-9 December 1994.

[10] B.Kao, H.Garcia-Molina, "Deadline assignment in a distributed soft realtime systems", Proc.of the 13th International Conference in Distributed Computing Systems, pp.428-437, 1993.

[11] A. Sahoo, W. Zaho, "Partition-based admission control in heterogeneous networks for hard real time connections", Proc. of the 10th International Conference on Parallel and Distributed Computing, October1997.

[12] T. Bonald, A. Poutière, J. Roberts, "Statistical performance guarantees for streaming flows using expedited forwarding", Proc. of IEEE INFOCOM'2001, March 2001.

[13] H. Takagi, "Queueing analysis, volume 1: vacation and priority systems", North Holland, Amsterdam, 1991.

A Quantitative QoS Routing Model for Diffserv Aware MPLS Networks

Haci A. Mantar[1,2]

[1] Department of Computer Engineering, Gebze Institute of Technology, Turkey
[2] Department of Computer Computer Engineering, Harran University, Turkey

Abstract. The paper proposes a pre-established multi-path model for quantitative QoS guarantees in Differentiated Services (Diffserv) aware MPLS networks. The proposed model pre-establishes several MPLS label switching paths (LSPs) between each ingress-egress router pair. An ingress router performs admission control based on the resource availability on these paths. The model reduces QoS route computation complexity and increases signaling and state scalability. It also increases resource utilization by performing dynamic load-balancing among the paths based on their utilization. The experimental results are provided to illustrate the efficiency of our model under various network conditions.

1 Introduction

Providing quantitative Quality of Services (QoS) in the Internet has become one of the most important issues as the use of real-time applications continue to grow dramatically. A QoS mechanism has two main components; QoS routing and resource reservation.

Most of the QoS routing schemes rely on source routing. In source routing model, each source node must have global QoS state information of all the nodes in the network in order to perform QoS routing. The global state information is updated by a link state protocol that requires a periodic exchange of link QoS state information among routers.

Source routing has several problems. First, under dynamic traffic conditions it is difficult to maintain accurate network state information in a scalable manner. Because network resource availability may change with each flow arrival and departure, and therefore a high frequent update is required to maintain accurate network state. However, high frequent updates result in communication and processing scalability problems. Second, the global view of the network QoS state may have synchronization problem. Third, performing QoS routing for each request can cause a serious scalability problem, because QoS routing is a computationally intensive task. Fourth, an ingress router may find a different path for each request. Since each computed route/path needs a reservation state in routers (i.e., for guaranteed QoS), the number of reservation states may increase accordingly with the number of accepted requests, resulting in a state scalability problem in the network core.

G. Parr, D. Malone, and M. Ó Foghlú (Eds.): IPOM 2006, LNCS 4268, pp. 61–71, 2006.
© Springer-Verlag Berlin Heidelberg 2006

As a viable alternative to source routing and its associated link state protocol, several studies have proposed a pre-established path approach [9][10][13][16].
In [9], the authors used a pre-computed multi-path approach that requires no
global state information exchanged among routers. Instead, source nodes obtain
the network QoS state based on flow blocking statistics collected locally, and performs flow routing using this localized view of the network QoS state. Elwalid et
al. [10] proposed a pre-established multi-path scheme for MPLS traffic engineering. The ingress router distributes the incoming traffic across pre-established
label switching paths (LSPs) based on the measurement of LSP congestion so
that the loads are balanced and congestion is thus minimized. Ref. [13] and [16]
proposed a Bandwidth Broker (BB) [1] based pre-established path scheme.

The pre-computed/established path schemes have several advantages. First of
all, since there is no need for global information exchange among the routers, the
communication overhead is minimized. Core routers do need to keep and update
any QoS state of other routers. Thus, it reduces the processing and memory
overhead at core routers. Another important issue is that routers do not perform
QoS route computation, which is a computationally extensive task. However,
these studies have not addressed *quantitative* QoS guarantees, because there are
no explicit admission control and reservation mechanisms. Since all the incoming
traffic is accepted, the QoS may be degraded as the traffic rate in the network
increases. Although Ref. [13] and [16] proposed an admission control scheme,
they used a single path scheme, which results in poor resource utilization. Also
as we showed in [16], performing QoS routing through bandwidth broker can
result in scalability problem.

In this work we propose a pre-established multi-path model for *quantitative*
QoS guarantees in Diffserv[5] over MPLS networks with an explicit admission
control scheme. In the proposed model, several LSPs, each of which has a certain
bandwidth constraint, are established for each ingress-ingress pair in offline. The
ingress router performs admission control based on the resource availability in
the associated LSPs and distributes the accepted traffic across the LSPs in a way
load is balanced and resource utilization is thus optimized. Another important
feature of our model is that since the Internet traffic behaves highly variant–
depending on applications, users, mobile devices etc, in the proposed model,
ingress routers dynamically readjust the size of LSPs with respect to LSPs'
traffic characteristics.

The remainder of the paper is organized as follows: Section II describes a
quantitative Diffserv model. Section III presents the QoS routing model and admission control model. Section IV describes the ingress equal-cost load balancing
scheme. In Section V, we present the preliminary evaluation results. Finally, the
conclusion is given in Section VI.

2   Quantitative Diffserv Network Model and Assumptions

Diffserv is seen to be a *de facto* standard to achieve QoS in the Internet in a scalable manner. Unlike Integrated Services (Intserv), Diffserv requires no per-flow

admission control or signaling and, consequently, core routers do not maintain any per-flow state or operation. Instead, routers merely implement a small number of classes named Per Hop Behavior(PHB), each of which has particular scheduling and buffering mechanisms. However, with the exception of Expedited Forwarding (EF) [7], all the PHBs provide *qualitative* QoS guarantees (the QoS metrics value changes with network conditions). Hence, the requirements of real-time applications, which need *quantitative* bounds on specific QoS metrics, cannot be guaranteed even in a single node. It is envisioned that the EF service can only be applied to very limited applications, because it results in poor resource utilization [7][11][12]. In this work, we define a set of PHBs that provide quantitative QoS guarantees.

A quantitative PHB i is associated with an upper delay bound d_i, an upper loss ratio bound l_i, and certain percentage of link capacity C_i. d_i is the delay that packets of i will experience in any router within the domain under the worst-case network conditions. l_i is the average loss rate that the packets of i will experience under worst-case network conditions (e.g., $d_i < 3ms$, $l_i < 0.01\%$). We assume that each PHB can use only its share of link capacity, and the surplus capacity of a PHB can be used by best-effort or qualitative services. We also assume that d_i and l_i are pre-determined at the network configuration stage [6][11][12], which is done in relatively long time intervals (e.g., weeks) and downloaded into routers. A router dynamically adjusts its scheduler rate and buffer size according to the dynamic traffic rate to meet pre-determined d_i and l_i constraints.

Under this premise, each router provides the desired QoS regardless of the utilization rate. This, of course, require a strict admission control mechanism to ensure that the incoming traffic rate of a PHB does not exceed the given link capacity for that particular PHB.

Note that since the link resources allocated for a PHB can only be used by that PHB, the network can be considered as if it is divided into multiple virtual networks, one for each PHB. Thus, in the rest of this paper, it is assumed that there is only one PHB within a domain. Consequently, the traffic of an LSP belongs to a single PHB (class).

3   QoS Routing and Admission Control

We propose a pre-established multi-path QoS routing model that minimizes the scalability problems while increasing resource utilization. We assume that several explicit paths, MPLS LSPs, have been pre-established between each ingress router (IR) and egress router (ER) pair with a certain bandwidth capacity (Figure 1). The traffic between an IR-ER pair is forwarded along these LSPs. For simplicity, at this point we assume that LSPs are PHB-specific, meaning that all the traffic in an LSP belongs to the same PHB (an LSP is defined with <*IR, ER, PHB*>).

In this work we do not address the LSP placement optimization, we simply assume that they have been pre-established by standard protocols such as RSVP-TE or CR-CDP. The LSPs placement is performed in network configuration

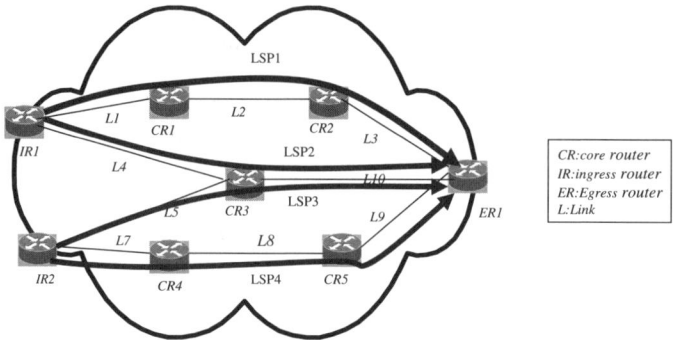

Fig. 1. A pre-established multi-path QoS routing architecture

stage, which is done in long time periods (e.g., daily, weekly)[9][10][16]. The initial capacity of an LSP can be chosen randomly, or it can be determined based on the traffic rate between an IR-ER pair. Since the size of LSPs are dynamically modified based on the traffic rate, their initial sizes do not affect the resource optimization.

The LSPs in our model are associated with certain QoS guarantees. The traffic entering an LSP is statistically guaranteed to be delivered to the other end of the LSP (egress router) within specified QoS boundaries, regardless of the network congestion level. This, of course, requires an explicit admission control mechanism. The admission control and resource reservation are performed by the ingress router. Upon receiving a QoS request:

- The ingress router determines the associated egress router (based on the destination address IP prefix).
- The ingress router checks the total available bandwidth in the associated LSPs (as described above there are several LSPs between an IR-ER pair). If the total available bandwidth is larger than the requested bandwidth, it accepts the request immediately without signaling the core and egress nodes. This minimizes admission control time and increases signaling scalability.
- If the available bandwidth is less than the requested bandwidth, the ingress router sends a resource allocation request (RAR) to egress nodes through the LSPs.
- All the nodes along the LSP process RAR and check if they have available resources to support the request. If so, they send the RAR to the next node. Otherwise, they send resource allocation answer (RAA) through the ingress router reporting insufficient resources (rejecting the request).
- When RAR reaches egress router, meaning that all the nodes between ingress and egress router have sufficient resources, the egress router reserves the required resources and send RAA to the ingress router.
- All the nodes along the path process RAA and reserves resources (increases the size of the associated LSP). Once the RAA reaches to the ingress router, the request is admitted.

Resizing LSPs for each request (sending RAR for individual each request) may result in a scalability problem (especially in the network core). To reduce this problem, the ingress router resizes LSPs based on the aggregated traffic demand in advance. That is, when the traffic rate between an IR-ER pair reaches the total capacity of LSPs, the ingress router attempts to increase the capacity of LSPs by taking the future request into account. For example, the size of each LSP can be increased by 10%. Similarly, when the traffic rate drop to below certain threshold (e.g., 80% of total capacity),the ingress router attempts to release some resources by reducing the size of the associated LSPs. It is important to note that reducing and increasing the sizes of LSPs is under the assumption that all the LSPs have equal cost. As described later, each LSP has a cost that reflects the utilization of the its links. The ingress router tries to equalize the costs of LSPs so that the load is balanced and congestion is thus minimized. The ingress router performs load balancing by shifting the traffic from high cost LSPs to low cost LSPs.

Another important scalability issue is the number of states maintained by core routers. In the proposed model, core routers do not maintain reservation states (neither per-flow nor per-LSP). Although each LSP has reserved bandwidth, this information is known only by the ingress router, which performs admission control. All the traffic of LSPs that use the same link and has the same PHB is aggregated into the same queue. A core router just ensures that the total aggregated traffic does not exceed its link capacity. When it accepts an RAR, it simply subtracts the requested bandwidth amount from the available one. It does not maintain state for the LSP size. The only states maintained by a core router are the forwarding states (associated with LSPs) and the total reserved capacity.

4   Ingress Equal-Cost Load Balancing

Given the traffic load flowing between an ingress-egress pair, the task of an ingress router is to distribute the traffic across the candidate LSPs in a way the loads are balanced and congestion is minimized. The idea here is to shift some of the traffic from the heavily-loaded path(s) to the lightly-loaded path(s) to increase admission control probability. For example; consider a scenario in Figure 1. Assume that LSP1, LSP2, and LSP3 have no available bandwidth left, and LSP4 has available bandwidth. Also assume that link 10 (L10) is the bottleneck link for LSP2 and LSP3. If IR1 receives new requests, it will reject them due to the lack of resources (no available bandwidth in LSP1 and LSP2). When the proposed scheme is used, some of the traffic in LSP3 will be shifted to LSP4 so that LSP2 can use more resources of L10. In this case, IR1 can increase the size of LSP2 and thus accept new requests. This is achieved through the cost-equalization of the LSPs.

The cost of a link for a PHB x can be simply defined[17] as

$$c_x(t) = q_x/(1 - u_x(t)) \tag{1}$$

where q_x is the fixed cost of using the link for x when it is idle, and $u_x(t)$ represents the link utilization of x at time t. $u_x(t) = r_x(t)/R_x$, where $r_x(t)$ is the traffic rate of x at time t and R_x represents the link capacity assigned to x.

The ingress router periodically receives the cost for each of the links that constituting its LSPs. Let $L1, L2, L3, ..LN$ be the links on an LSP i. Once the ingress router has the cost of all links, it computes the cost of i as

$$Ci = \sum_{j=1}^{N} c_j \qquad (2)$$

Upon detecting a consistent and substantial cost difference among LSPs, the ingress router invokes the load balancing/cost-equalization algorithm. The cost equalization can be done using stochastic approximation theory or gradient projection methods[10][15], which are generally complex to implement in practice. To circumvent this problem, we use a simple but effective iterative procedure.

Let $C_1, C_2, ..., C_K$ be the costs, $R_1, R_2, ..., R_k$ be the traffic rate and $\alpha_1, \alpha_2, ...,$ α_K be load proportions of $LSP_1, LSP_2, ..., LSP_K$ and V the be total load of an IR-ER pair ($\sum_{i=1}^{K} \alpha_i = 1$ and $R_i = \alpha_i V$) . If all C_i are equal, then α_i's are the desired proportions. If not, we use mean costs of all the paths $C^- = \sum C_i / K$ as the target cost for each path and obtain new proportions. The new proportions α_i^{n}'

$$\alpha_i^{'} = \frac{C^-}{C_i} \alpha_i \qquad (3)$$

$$\alpha_i^{n} = \phi \alpha_i^{'} \qquad (4)$$

ϕ is the normalization factor defined as $\phi = \frac{1}{\sum_{i=1}^{K} \alpha_i^{'}}$. The corresponding R_i^{n}'s are: $R_i^{n} = \alpha_i^{n} V$. This procedure is repeated iteratively until the costs of LSPs are equal. Because for a constant LSP size (during the load balancing) the cost C_i is increased with its load ($\alpha_i V$), it can be shown that this procedure will always converge.

For clarification, consider Figure 1. Let $C1, C2, C3$, and $C4$ be the costs and $R1, R2, R3$ and $R4$ be the traffic rate of LSP1, LSP2, LSP3 and LSP4, respectively. Suppose that the ingress routers IR1 and IR2 detect substantial cost differences between their LSPs, $C2 > C1$ and $C3 > C4$. In this situation, both IR1 and IR2 start shifting some traffic from LSP2 to LSP1 and from LSP3 to LSP4, respectively. That is, $R1, R2, R3$ and $R4$ will become R1+$\triangle$r1, R2-$\triangle$r1, R3-$\triangle$r2, and R4+$\triangle$r2, respectively ($\triangle r = R_i^{n} - R_i$). This process is repeated until there is no appreciable cost difference.

Another important issue is how to assign the load to the LSPs. There are at least two ways to do this: Packet-based in a round-robin fashion and flow-based using hashing on source and destination IP addresses and possibly other fields of the IP header. The first approach causes out-of-order packets while the second approach is dependent upon the distribution of flows. Flow-based load sharing may be unpredictable in a stub network where the number of flows is relatively small and less heterogeneous, but it is generally effective in core public networks where the number of flows is large and heterogeneous [2]. We therefore use the second approach.

In our hashing model, the traffic will be first distributed into N bins by using module-N operation on the hash space [8]. The router performs a modulo-N hash over the packet header fields that identify a flow. If the total traffic rate is X bps, each bin approximately receives the amount of X/N bps. The next step is to map N bins to LSPs. The number of bins assigned to an LSP is determined based on its load portion.

5   Experimental Results

In this section, we present experimental results to verify the achievements of the proposed model. We modified our previous software tools [2][16] for this purpose. For load-balancing experiments, we used a network topology consisting 8 edge routers (ingress, egress) and 8 core routers (CR) (Figure 2). For each IR-ER pair, two LSP were established, so there were 32 LSPs. Although an IR had LSP connection to all the ERs, in the figure the connection to a single ER is shown (for simplicity illustration).

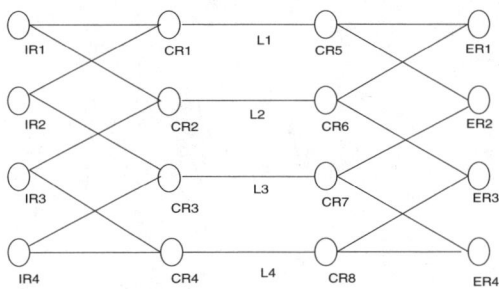

Fig. 2. Experiment Network Topology for load balancing

Figure 3 and 4 show, the load-balancing (cost-equalization) process for cost update interval of 1sec and 5sec. In this experiment, the offered traffic rate between each IR-ER pair was changed every 60 sec. The figures shows utilization of L1, L2 and L3 under different traffic conditions. As figures shows, our model achieves load-balancing in few iterations.

In the next experiments, we evaluate the model in terms of network resource utilization and admission control time on the network topology shown on Figure 5. Each ingress router had 3 LSPs to egress router (node 14). We compared our model with traditional per-request QoS routing scheme, which is also known as widest shortest path (wsp) in the literature. Per-request routing performs route computation for each individual request/flow. It searches for a feasible path with minimum hop count. If there are several such paths, the one with maximum available bandwidth is chosen. As known, unlike our model, in per-request routing the sessions/flows are not rerouted, meaning that there is no shifting process.

Figure 6 compares our model with per-request routing in terms of resource utilization. In this experiment, the ingress router attempted to increase the size

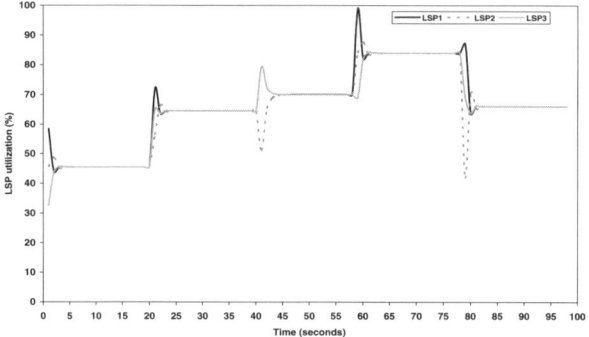

Fig. 3. The LSP cost-equalization (1 sec intervals)

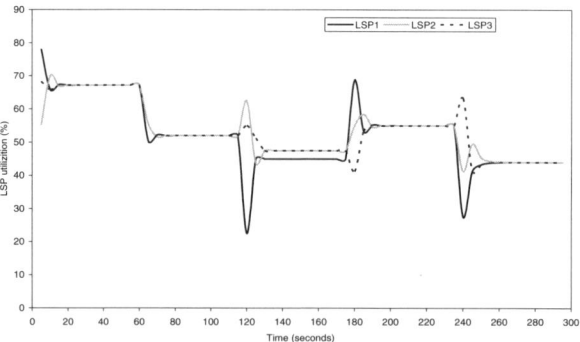

Fig. 4. The LSP cost-equalization (5 sec intervals)

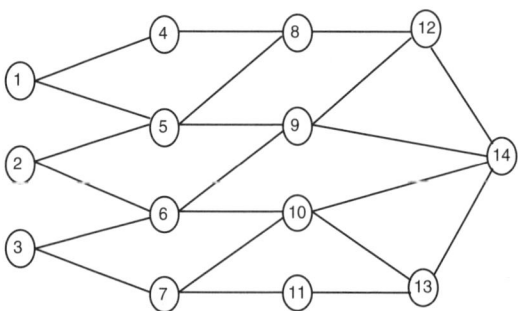

Fig. 5. Experiment Network Topology for admission control and resource utilization

of LSPs when the total traffic rate reaches 95% of the reserved capacity and de-crease when the total traffic rate drop to below 85%. Increasing and decreasing amount is 5% of the LSP capacity. As shown, under low loads, the difference in the rejection rates are relatively small because the probability of finding sufficient available bandwidth in each link is high and most of the requests are accepted. When the loads increases, our model has better performance than per-request routing. An important point here is that our model uses bulk-type reservation, meaning that an ingress router can increase LSP sizes more than the actual request rate (in order to damp the signaling frequency). This affects the performance of our model. As shown in Figure 7, when we reserve resources for each individual request, the performance of our model gets even much better because there is no over-reservation.

As described before, upon receiving a reservation request, the ingress router performs the admission control based only on the current utilization of the corresponding LSPs. It does not have to check the capacity of the links constituting the LSPs in online. In per-request QoS routing (wsp), the ingress router first computes the path and then required resources needs to reserved in all the links along the path. Figure 8 and Figure 9 illustrate the performance of our model with per-request routing in terms admission control time.

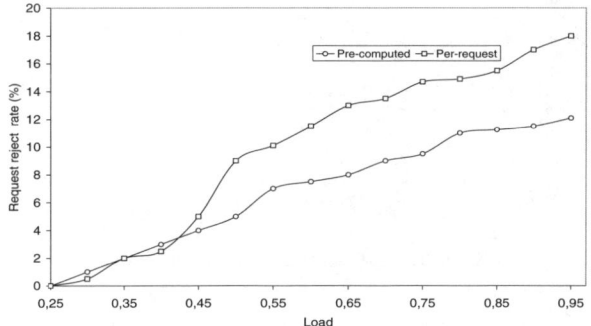

Fig. 6. Resource utilization with over-reservation

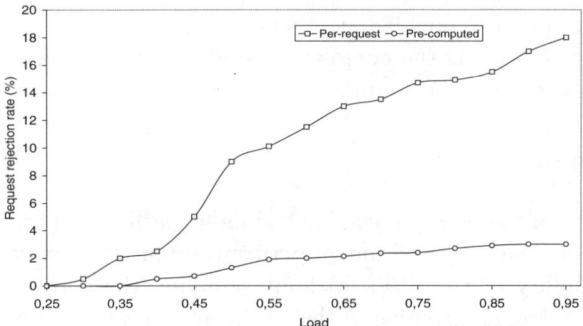

Fig. 7. Resource utilization without over-reservation

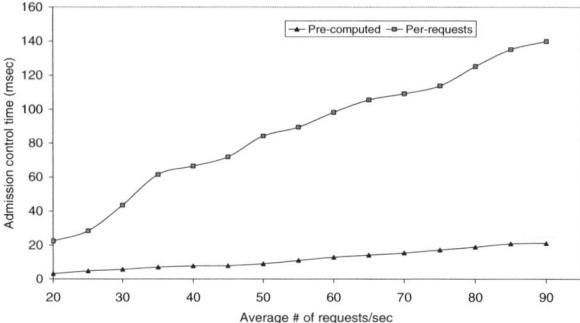

Fig. 8. The admission control time

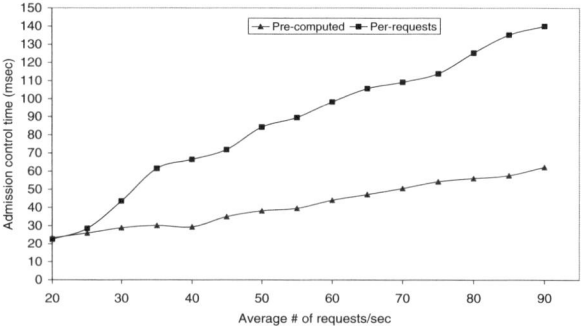

Fig. 9. The admission control time

Figure 8 shows the results when the resources of LSPs were reserved in advance with over-utilization. As expected, admission control time of our model is very small compare to per-request QoS routing. In Figure 9, the ingress router reserved the resources on LSPs when a request arrived. In other words, resources were reserved and released based on the requests. As shown, even in this case, the performance of our model is much better than per-request QoS routing. Although we do not provide the complexity analysis in this work, the admission control time also reflects the complexity.

6   Conclusion

In this paper we proposed a pre-established multi path model for Diffserv aware MPLS networks. The proposed model performs admission control based on the resource availability of the LSPs without signaling the core node. The loads among paths are balanced based on their utilization costs dynamically obtained via measurements. The model significantly reduces admission control time and minimizes scalability problems present in prior research while optimizing network resource utilization.

References

1. K. Nichols, V. Jacobson, and L. Zhang. " A Two-bit Differentiated Services Architecture for the Internet" RFC 2638, July 1999.
2. H. Mantar, J.Hwang, S. Chapin, I. Okumus "A Scalable Model for Inter-Bandwidth Broker Resource Reservation and Provisioning", IEEE Journal on Selected Areas in Communications (JSAC), Vol.22, No.10, December 2004
3. QBone Signaling Design Team, "Simple Inter-domain Bandwidth Broker Signaling (SIBBS)", http://qbone.internet2.edu/bb/ work in progress.
4. E. Rosen, A. Viswanathan, R. Callon "Multiprotocol Label Switching Architecture", RFC 3031.
5. S. Black et al., "An Architecture for Differentiated Services," RFC2475, Dec. 1998.
6. K. Nichols, B. Carpenter "Definition of Differentiated Services Per Domain Behaviors and Rules for their Specification" RFC 3086.
7. V. Jacobson, K. Nichols, K. Poduri, "An Expedited Forwarding PHB," RFC2598, June 1999.
8. C. Hopps "Analysis of an Equal-Cost Multi-Path Algorithm" RFC 2992, November 2000
9. S. Nelakuditi, Z.-L. Zhang, R.P. Tsang, and D.H.C. Du, Adaptive Proportional Routing: A Localized QoS Routing Approach, IEEE/ACM Transactions on Networking, December 2002.
10. A. Elwalid, C. Jin, S. Low, and I. Widjaja, "MATE: MPLS Adaptive Traffic Engineering", IEEE INFOCOM 2001
11. N. Christin, J. Liebeherr and T. Abdelzaher. A Quantitative Assured Forwarding Service. In Proceedings of IEEE INFOCOM 2002.
12. S. Wang, D. Xuan, R. Bettati, and W. Zhao, "Providing Absolute Differentiated Services for Real-Time Applications in Static-Priority Scheduling Networks," in IEEE/ACM Transactions on Networking, Vol. 12, No. 2, pp. 326-339, April 2004.
13. Z.-L. Zhang, Z. Duan, L. Gao, and Y. T. Hou, "Decoupling QoS control from core routers: A novel bandwidth broker architecture for scalable support of guaranteed services", In Proc. ACM SIGCOMM, August 2003
14. T.Li and Y. Rekhter."A provider Architecture for Differentiated services and Traffic Engineering RFC2490
15. D. Bertsekas, R. Gallager, "Data networks," 2nd edition, PrenticeHall, Inc., 1992
16. H. Mantar, I. Okumus, J. Hwang, S. Chapin " An Intra-domain Resource Management Model for Diffserv Networks", Journal of High Speed Networks, Vol. 15, pp. 185205, April 2006
17. Haci A. Mantar, "An Admission Control and Traffic Engineering Model for Diffserv-MPLS Networks," to appear on APPNOMS, Sept. 2006, Buson, Korea

Experience-Based Admission Control with Type-Specific Overbooking*

Jens Milbrandt, Michael Menth, and Jan Junker

University of Würzburg, Institute of Computer Science, Germany
{milbrandt, menth, junker}@informatik.uni-wuerzburg.de

Abstract. Experience-based admission control (EBAC) is a hybrid approach combining the classical parameter-based and measurement-based admission control schemes. EBAC calculates an appropriate overbooking factor used to overbook link capacities with resource reservations in packet-based networks. This overbooking factor correlates with the average peak-to-mean rate ratio of all admitted traffic flows on the link. So far, a single overbooking factor is calculated for the entire traffic aggregate. In this paper, we propose type-specific EBAC which provides a compound overbooking factor considering different types of traffic that subsume flows with similar peak-to-mean rate ratios. The concept can be well implemented since it does not require type-specific traffic measurements. We give a proof of concept for this extension and compare it with the conventional EBAC approach. We show that EBAC with type-specific overbooking leads to better resource utilization under normal conditions and to faster response times for changing traffic mixes.

Keywords: admission control, resource reservation overbooking, quality of service, traffic management & control.

1 Introduction

Admission control (AC) may be used to ensure quality of service (QoS) in terms of packet loss and delay in packet-based communication networks. Many different approaches for AC exist and an overview can be found in [1]. In general, AC admits or rejects resource reservation requests and installs reservations for admitted flows. The packets of admitted flows are transported with high priority such that they get the desired QoS. Rejected flows are either blocked or their packets are handled only with lower priority.

Link admission control (LAC) methods protect a single link against traffic overload. They can be further subdivided into parameter-based AC (PBAC) and measurement-based AC (MBAC). PBAC methods [2, 3, 4] use traffic descriptors to calculate a priori the expected bandwidth consumptions of admitted flows to get an estimate of the remaining free capacity which is required for future admission decisions. PBAC offers stringent QoS guarantees to data traffic that has been admitted to the network, but it lacks scalability with regard to the signalling of resource reservations. In contrast, there

---

* This work was funded by Siemens AG, Munich, and the German Research Foundation (DFG).The authors alone are responsible for the content of the paper.

G. Parr, D. Malone, and M. Ó Foghlú (Eds.): IPOM 2006, LNCS 4268, pp. 72–83, 2006.

are numerous measurement-based AC (MBAC) approaches which use real-time measurements to assess the remaining free capacity [5, 6, 7, 8, 9, 10, 11, 12, 13]. MBAC uses the available network resources very efficiently, but it relies on real-time traffic measurements and, therefore, it is susceptible to QoS violation.

Experience-based admission control (EBAC) is a hybrid solution [14]. It uses peak rate allocation based on traffic descriptors and calculates a factor to overbook a given link capacity. The calculation of this overbooking factor is based on the statistics of the utilization of past reservations that are obtained by measurements. Hence, EBAC does not require real-time measurements of the instantaneous traffic for admission decisions and is, therefore, substantially different from classical MBAC approaches and easier to implement. The major task of EBAC is the calculation of an appropriate overbooking factor for classical PBAC. This factor is obtained by measurements and correlates with the average peak-to-mean rate ratio (PMRR) of all admitted flows which only indicate their peak rate. In previous work, we have provided a proof of concept for EBAC [15]. We also investigated its robustness during sudden changes of the traffic properties to which all MBAC methods are susceptible [16]. So far, a single overbooking factor is calculated based on the traffic characteristics of the entire admitted traffic aggregate. This paper extends EBAC towards type-specific overbooking (TSOB) which provides a compound overbooking factor considering different types of traffic. The extension can be well implemented since it does not require type-specific measurements. We give a proof of concept for EBAC with TSOB and compare it with the conventional EBAC approach. We show that EBAC with TSOB leads to better resource utilization under normal traffic conditions and to faster response times in case of changing traffic mixes. Unlike conventional EBAC, the extension avoids congestion due to overreservation if the fraction of flows with low PMRR increases in the traffic mix. All of the above sketched AC mechanisms apply for a single link, but they can be extended on a link-by-link basis for a network-wide application. For the sake of clarity, we limit our performance study to a single link which can be done without loss of generality.

This paper is structured as follows. In Section 2, we briefly review the EBAC concept. Section 3 describes our simulation design and the applied traffic model and summarizes results from previous studies. Section 4 proposes the extension of EBAC towards type-specifc overbooking (TSOB). The simulation results in Section 5 show the superiority of EBAC with TSOB over conventional EBAC. Finally, Section 6 summarizes this work and points out further steps towards the application of type-specific overbooking in practice.

2 Experience-Based Admission Control (EBAC)

In this section, we briefly review the EBAC concept with emphasis on the EBAC memory which implements the experience based on which AC decisions are made.

An AC entity limits the access to a link l with capacity $c(l)$ and records all admitted flows $f \in \mathcal{F}(t)$ at any time t together with their requested peak rates $\{r(f) : f \in \mathcal{F}(t)\}$. When a new flow f_{new} arrives, it requests a reservation for its peak rate $r(f_{new})$. If

$$r(f_{new}) + \sum_{f \in \mathcal{F}(t)} r(f) \leq c(l) \cdot \varphi(t) \cdot \rho_{max} \tag{1}$$

holds, admission is granted and f_{new} joins $\mathcal{F}(t)$. If flows terminate, they are removed from $\mathcal{F}(t)$. For conventional PBAC systems, the overbooking factor is $\varphi(t)=1$ while for EBAC, the experience-based overbooking factor $\varphi(t)$ is calculated by statistical analysis and indicates how much more bandwidth than $c(l)$ can be safely allocated for reservations. The maximum link utilization threshold ρ_{max} limits the traffic admission such that the expected packet delay W exceeds an upper delay threshold W_{max} only with probability p_W. We calculate the threshold ρ_{max} based on the $N \cdot D/D/1 - \infty$ approach [17].

For the computation of the overbooking factor $\varphi(t)$, we calculate the time-dependent reserved bandwidth of all flows by $R(t) = \sum_{f \in \mathcal{F}(t)} r(f)$. EBAC performs traffic measurements $M(t)$ on the link and collects a time statistic for the reservation utilization $U(t) = M(t)/R(t)$. The value $U_p(t)$ denotes the p_u-percentile of the empirical distribution of U and the reciprocal of this percentile is the overbooking factor $\varphi(t) = 1/U_p(t)$.

The EBAC system requires a set of functional components to calculate the overbooking factor $\varphi(t)$:

1. **Measurement Process for $M(t)$** — To obtain $M(t)$, we use disjoint interval measurements such that for a time interval I_i with length Δ_i, the measured rate $M_i = \Gamma_i/\Delta_i$ is determined by metering the traffic volume Γ_i sent during I_i.

2. **Statistic Collection $P(t,U)$** — For the values $R(t)$ and $M(t)$, a time statistic for the reservation utilization $U(t) = M(t)/R(t)$ is collected. The values $U(t)$ are sampled in constant time intervals and are stored as hits in bins for a time-dependent histogram $P(t,U)$. From this histogram, the time-dependent p_u-percentile $U_p(t)$ of the empirical distribution of U can be derived as

$$U_p(t) = \min_u \{u : P(t, U \leq u) \geq p_u\}. \qquad (2)$$

3. **Statistic Aging Process for $P(t,U)$** — If traffic characteristics change over time, the reservation utilization statistic must forget obsolete data to reflect the properties of the new traffic mix. Therefore, we record new samples of $U(t)$ by incrementing the corresponding histogram bins by one and devaluate the contents of all histogram bins in regular devaluation intervals I_d by a constant devaluation factor f_d. The devaluation process determines the memory of EBAC which is defined next.

4. **Memory of EBAC** — The histogram $P(t,U)$, i.e. the collection and the aging of statistical AC data, is the memory of EBAC. This memory correlates successive flow admission decisions and influences the adaptation of the overbooking factor $\varphi(t)$ in case of traffic changes on the link. The statistic aging process is characterized by the devaluation interval I_d and the devaluation factor f_d. It makes the memory forget about reservation utilizations in the past. The parameter pairs (I_d, f_d) yield typical half-life periods T_H after which collected values $U(t)$ have lost half of their importance in the histogram. Therefore, we have $\frac{1}{2} = f_d^{T_H/I_d}$ and define the EBAC memory based on its half-life period

$$T_H(I_d, f_d) = I_d \cdot \frac{-ln(2)}{ln(f_d)}. \qquad (3)$$

3   EBAC Performance Simulation

In this section, we first present the simulation design of EBAC on a single link and the traffic model we used on the flow and packet scale level. Afterwards, we summarize recent EBAC simulation results from [15, 16].

3.1   Simulation Design

The design of our simulation is shown in Figure 1. Different types of traffic *source generators* produce flow requests that are admitted or rejected by the *admission control* entity. To make an admission decision, this entity takes the overbooking factor $\varphi(t)$ into account. In turn, it provides information regarding the reservations $R(t)$ to the *EBAC system* and yields flow blocking prababilities $p_b(t)$. For each admitted source, a *traffic generator* is instantiated to produce a packet flow that is shaped to its contractually defined peak rate. Traffic flows leaving the *traffic shapers* are then multiplexed on the buffered link with capacity $c(l)$. The link provides information regarding the measured traffic $M(t)$ to the EBAC system and yields packet delay probabilities $p_d(t)$ and packet loss probabilities $p_l(t)$. Another measure for the performance of EBAC is the overall response time T_R, i.e., the time span required by the EBAC system to adapt the over-booking factor to a new traffic situation. The time T_R depends on the transient behavior of EBAC and is investigated in [16].

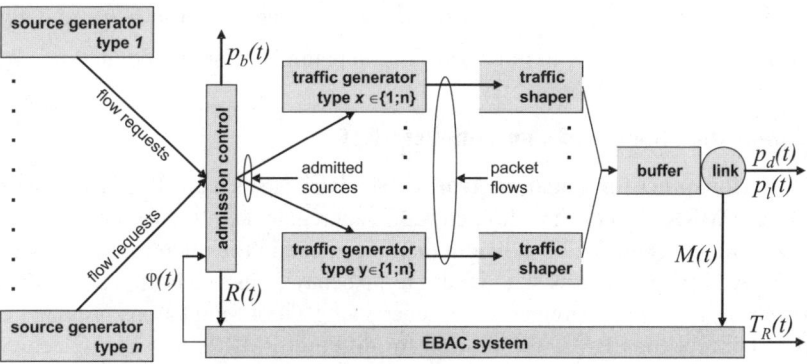

Fig. 1. Simulation design for EBAC in steady and transient state

3.2   Traffic Model

In our simulations, the traffic controlled by EBAC is modelled on a flow scale level and a packet scale level. While the flow level controls the inter-arrival times of flow requests and the holding times of admitted flows, the packet level defines the inter-arrival times and the sizes of packets within a single flow.

Flow Level Model. On the flow level, we distinguish different traffic source types, each associated with a characteristic peak-to-mean rate ratio (PMRR) and corresponding to a source generator type in Figure 1. The inter-arrival time of flow requests and the holding time of admitted flows both follow a Poisson model [18], i.e., new flows arrive with rate λ_f and the duration of a flow is controlled by rate μ_f. The mean of the flow inter-arrival time is thus denoted by $1/\lambda_f$ and the holding time of a flow is exponentially distributed with a mean of $1/\mu_f$. Provided that no blocking occurs, the overall offered load $a_f = \lambda_f/\mu_f$ is the average number of simultaneously active flows measured in Erlang. We analyze the EBAC with a load of $a_f \geq 1.0$, i.e., we consider high load scenarios where the link is mostly saturated with reservations which is a prerequisite to make the effect of AC visible.

Packet Level Model. We use a rather simple parameterizable packet level model instead of real traffic traces because we conduct simulations where we want to control the properties of the flows. We use a fixed packet size and assume that the inter-arrival time of the packets is distributed exponentially with a mean rate $c(f)$. We are aware of the fact that Poisson is not a suitable model to simulate Internet traffic on the packet level [19]. Therefore, we shape consecutive packets according to a certain peak rate $r(f)$ (cf. Figure 1) which influences the flow properties significantly.

In practice, applications know and signal the peak rates $r(f)$ of their corresponding traffic flows. The type of an application can be determined, e.g., by a signalling protocol number. We use only this limited information in our simulations, i.e., the mean rates $c(f)$ of the flows are not known to the EBAC measurement process, they are just model parameters for the traffic generation. Therefore, we can control the rate of flow f by its peak-to-mean rate ratio (PMRR) $k = \frac{r(f)}{c(f)}$. The mean rate of the admitted traffic aggregate $C(t) = \sum_{f \in \mathcal{F}(t)} c(f)$ is also unknown in practice, but it helps to define its PMRR $K(t) = \frac{R(t)}{C(t)}$ which is an important control parameter for our simulation.

3.3   Simulation Studies of Conventional EBAC

EBAC Performance for Constant Traffic. The intrinsic idea of EBAC is the exploitation of the PMRR $K(t)$ of the admitted traffic aggregate, i.e., to take advantage of the fact that flows reserve more bandwidth than they need in the middle. In [15], we simulated EBAC on a single link with regard to its behavior in steady state, i.e., when the properties of the traffic aggregate were rather static. These simulations provided a first proof of concept for EBAC. We showed for different PMRRs that EBAC achieves a high degree of resource utilization through overbooking while packet loss and packet delay are well limited. The simulation results allowed us to give recommendations for the EBAC parameters such as measurement interval length and reservation utilization percentile to obtain appropriate overbooking factors $\varphi(t)$. They furthermore showed that the EBAC mechanism is robust against traffic variability in terms of packet size and inter-arrival time distribution as well as against correlations thereof.

EBAC in the Presence of Traffic Changes. As EBAC partly relies on traffic measurements, it is susceptible to changes of the traffic characteristics of admitted flows with regard to QoS because individual flows can suddenly send with their peak rate even though they used to send less traffic before. We briefly summarize the results from [16] where we investigated the transient behavior of conventional EBAC after sudden traffic changes. On the one hand, the performance measures were the QoS performance in terms of packet loss $p_l(t)$ and packet delay $p_d(t)$ (cf. Figure 1) which are potentially compromised in case of suddenly increasing traffic rates (= decreasing PMRR). On the other hand, the duration from the sudden change of the PMRR to the time where the overbooking factor $\varphi(t)$ of the EBAC has adapted to the new PMRR is an interesting measure for the EBAC that we called its response time $T_R(t)$. The experiments investigated the performance of EBAC under very extreme traffic conditions that correspond to a collaborative and simultaneous QoS attack by all traffic sources. We showed that the response time T_R depends linearly on the half-life period T_H in case of a sudden change of the traffic intensity. For decreasing traffic intensity (= increasing PMRR) the QoS of the traffic is not at risk. However, for a suddenly increasing traffic intensity (= decreasing PMRR) the QoS is compromised for a certain time span.

4   EBAC with Type-Specific Overbooking

In this section, we present type-specific overbooking (TSOB) as a concept extending EBAC. So far, we only considered the peak-to-mean rate ratio (PMRR) of the entire admitted traffic aggregate and calculated a single factor to overbook the link capacity. We now include additional information about the characteristics of individual traffic types and their share in the currently admitted traffic mix to calculate a compound type-specific overbooking factor. First, we describe the system extension and then we show how the compound overbooking factor for EBAC with TSOB can be estimated without type-specific traffic measurements. Finally, we present some simulation results showing the advantage of EBAC with TSOB over conventional EBAC.

4.1   EBAC System Extension

We assume that different applications produce traffic flows with typical PMRRs that remain rather constant over time. This leads to different traffic types i ($1 \leq i \leq n$) that subsume flows with similar PMRRs from different applications. These traffic types have then characteristic utilization quantiles $U_{p,i}(t)$ and overbooking factors $\varphi_i(t) = 1/U_{p,i}(t)$. The share of traffic type i regarding all reservations is expressed by the value $\alpha_i(t) = R_i(t)/R(t)$ with $R(t) = \sum_{i=0}^{n} R_i(t)$. The shares of all traffic types is represented by the vector

$$\alpha(t) = \begin{pmatrix} \alpha_1(t) \\ \vdots \\ \alpha_n(t) \end{pmatrix} \text{ with } \sum_{i=1}^{n} \alpha_i(t) = 1. \tag{4}$$

EBAC with TSOB uses the information about the time-dependent traffic composition $\alpha(t)$ and the overall reservation utilization $U(t)$ to calculate the time-dependent type-specific reservation utilizations $U_i(t)$. Their estimation is a rather complex and described in Section 4.2. With type-specific measurements $M_i(t)$ and type-specific reservation utilizations $U_i(t) = \frac{M_i(t)}{R_i(t)}$, we have the relation

$$U(t) = \frac{M(t)}{R(t)} = \frac{\sum_{i=1}^{n} M_i(t)}{\sum_{i=1}^{n} U_i(t)} = \frac{\sum_{i=1}^{n} U_i(t) \cdot R_i(t)}{R(t)} = \sum_{i=1}^{n} \alpha_i(t) \cdot U_i(t). \tag{5}$$

The values $U_i(t)$ are stored as hits in bins of separate histograms $P_i(t,U)$ which yield type-specific reservation utilization percentiles $U_{p,i}(t)$. For EBAC with TSOB, the admission decision of the conventional EBAC in Equation (3) then extends to

$$r(f_i^{new}) \cdot U_{p,i}(t) + \sum_{f \in \mathcal{F}(t)} r(f) \cdot U_{p,type(f)}(t) \leq c(l) \cdot \rho_{max} \qquad (6)$$

for a new flow f_i^{new} of type i. Note that the general overbooking factor $\varphi(t)$ on the right side in Equation (3) is substituted by type-specific utilization quantiles $U_{p,i}(t)$ on the left side of this equation. Assuming that for the utilization quantiles holds the same robust relation as in Equation (5), we calculate the overall overbooking factor for EBAC with TSOB by

$$\varphi(t) = \frac{1}{\sum_{i=1}^{n} \alpha_i(t) \cdot U_{p,i}(t)} \qquad (7)$$

and use it in the performance study in Section 5.

4.2   Estimation of Type-Specific Reservation Utilizations

A crucial issue for the performance of EBAC with TSOB is the estimation of the type-specific reservation utilizations $U_i(t)$. Type-specific measurements $M_i(t)$ yield exact values for $U_i(t) = M_i(t)/R_i(t)$. For a reduced number of traffic classes, type-specific measurements seem feasible if we consider new network technologies such as differentiated services (DiffServ) [20] for traffic differentiation and multi protocol label switching (MPLS) [21] for the collection of traffic statistics. However, current routers mostly do not provide these type-specific traffic measurements.

In the following, we develop a method to obtain estimates for the type-specific reservation utilizations that uses only the available parameters $M(t)$, $R(t)$, $R_i(t)$, and $\alpha(t)$ to estimate the $U_i(t)$ and that does not require type-specific measurements $M_i(t)$. The approach is based on a least squares approximation (LSA, cf. e.g. [22]) of the values $U_i(t)$. We illustrate it for two different traffic types $i \in \{1,2\}$. $U_1(t)$ and $U_2(t)$ denote their type-specific reservation utilizations. The global reservation utilization is then $U(t) = \alpha_1(t) \cdot U_1(t) + \alpha_2(t) \cdot U_2(t)$ and with $\alpha_1(t) + \alpha_2(t) = 1$ we get

$$U(t) = \alpha_1(t) \cdot (U_1(t) - U_2(t)) + U_2(t). \qquad (8)$$

We substitute $a_j = U_1(t_j) - U_2(t_j)$ and $b_j = U_2(t_j)$ and obtain the least squares error for parameters $U_1(t)$ and $U_2(t)$ if we minimize the term

$$\mathcal{L} = \min_{a_m,b_m} \sum_{j=1}^{m} [U(t_j) - (\alpha_1(t_j) \cdot a_m + b_m)]^2. \qquad (9)$$

The time index j thereby covers all values $U(t_j)$ and $\alpha(t_j)$ from the first ($j = 1$) to the last ($j = m$) probe ever determined by the EBAC system. We find the minimum of $\mathcal{L}$ where the first derivatives of Equation (9) yield zero, i.e., we set $\frac{\partial \mathcal{L}}{\partial a} \overset{!}{=} 0$ und $\frac{\partial \mathcal{L}}{\partial b} \overset{!}{=} 0$ and resolve these equations to parameters a_m and b_m yielding

$$a_m = \frac{m \cdot \sum_j \alpha_1(t_j) U(t_j) - \sum_j \alpha_1(t_j) \cdot \sum_j U(t_j)}{m \cdot \sum_j \alpha_1(t_j)^2 - \left(\sum_j \alpha_1(t_j)\right)^2} \tag{10a}$$

$$b_m = \frac{\sum_j U(t_j) \cdot \sum_j \alpha_1(t_j)^2 - \sum_j \alpha_1(t_j) \cdot \sum_j \alpha_1(t_j) U(t_j)}{m \cdot \sum_j \alpha_1(t_j)^2 - \left(\sum_j \alpha_1(t_j)\right)^2} \tag{10b}$$

for $1 \leq j \leq m$. The sums in Equations (10a) and (10b) can be computed iteratively which helps to cope with the large set of instances observed over all times t_j. In addition, we apply the time exponentially weighted moving average (TEWMA) algorithm to these sums to blind out short-time fluctuations. Due to the lack of space, we omit any details of the TEWMA algorithm which is described in [23]. With the calculated parameters a_m and b_m, we finally obtain the type-specific reservation utilizations $U_1(t_m) = a_m + b_m$ and $U_2(t_m) = b_m$.

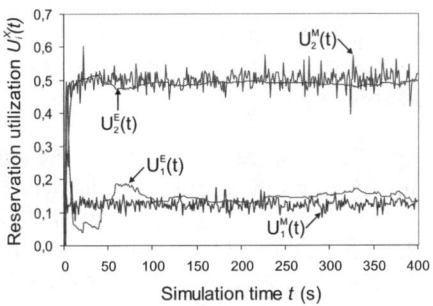

Fig. 2. Measured and estimated type-specific reservation utilizations

We perform simulations for estimating the type-specific reservation utilizations. Figure 2 shows a comparison of the measured type-specific reservation utilizations $U_i^M(t)$ and their corresponding estimates $U_i^{LSA}(t)$.

Our simulation contains two traffic types $i \in \{1, 2\}$. Type 1 has a PMRR $K_1 = 2$ and a mean share of $\alpha_1 = 0.2$ in the traffic mix. Type 2 has a PMRR $K_2 = 8$ and a mean share $\alpha_2 = 0.8$. All values K_i and α_i are averages. The type-specific reservation utilizations are determined every second. On the packet level, we have Poisson distributed inter-arrival times which lead to short-time fluctuations for the measured values $U_i^M(t)$. These fluctuations are clearly damped by the TEWMA algorithm used for the estimated values $U_i^{LSA}(t)$. The LSA provides good estimates for the corresponding measured values after some time. Hence, this estimation method enables EBAC with TSOB without type-specific traffic measurements.

5 Performance Comparison of Conventional EBAC and EBAC with TSOB

To investigate EBAC with TSOB, we perform a number of simulations each associated with a different traffic situation. For all simulations, we use a link capacity $c(l) = 10$ Mbit/s and simulate with two traffic types $i \in \{1, 2\}$ with characteristic peak-to-mean rate ratios (PMRRs) $K_1 = 2$ and $K_2 = 8$. A flow f_i of any type i reserves bandwidth with a peak rate $r(f_i) = 768$ Kbit/s and has a mean holding time of $1/\mu_f = 90$ s. The mean interarrival time of flow requests is set to $1/\lambda_f = 750$ ms such that the link is saturated with traffic, i.e., some flow requests are rejected. For conventional EBAC we use the overbooking factor according to Section 2 and for EBAC with TSOB, we calculate it according to Equation (7). In the following two simulation experiments, we focus on the

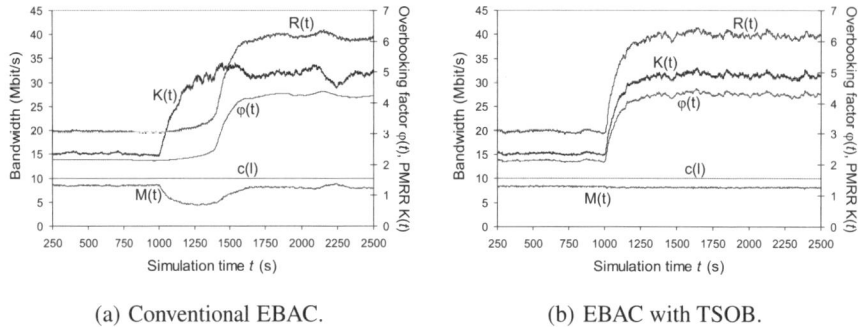

(a) Conventional EBAC.   (b) EBAC with TSOB.

Fig. 3. Conventional EBAC vs. EBAC with TSOB during a traffic intensity decrease

reaction of EBAC with TSOB after a decrease or an increase of the traffic intensity. We consider sudden changes of the traffic composition $\alpha(t)$ to have worst case scenarios and to obtain upper bounds on the EBAC response times.

Simulation with Decreasing Traffic Intensity. We investigate the change of the traffic intensity from a high to a low value. Figure 3 shows the average results over 50 simulation runs. We use the same two traffic types with their characteristic PMRRs as before. However, we start with mean traffic shares $\alpha_1 = 0.8$ and $\alpha_2 = 0.2$. At simulation time $t_0 = 1000$ s, the mean shares of both traffic types are swapped to $\alpha_1 = 0.2$ and $\alpha_2 = 0.8$ by changing the type-specific request arrival rates, i.e., the traffic intensity of the entire aggregate decreases due to a change in the traffic mix $\alpha(t)$. This leads to a sudden increase of the PMRR $K(t)$ which results in an immediate decrease of the measured traffic $M(t)$ for conventional EBAC (cf. Figure 3a). With observable delay, the conventional EBAC system adapts its overbooking factor $\varphi(t)$ as a result of the slowly decreasing p_u-percentile $U_p(t)$ in the histogram $P(t, U)$. From other simulations [16] we know that this delay strongly depends on the EBAC memory defined by the half-life period T_H in Equation (3). In contrast, EBAC with TSOB (cf. Figure 3b) increases its overbooking factor $\varphi(t)$ almost at once since the p_u-percentiles of the type-specific histograms $P_i(t, U)$ remain rather constant. As only the shares of the traffic types in the mix have changed, the compound $\varphi(t)$ is immediately adapted. As a consequence, the faster reaction of EBAC with TSOB leads to a higher and more stable mean link utilization.

Simulation with Increasing Traffic Intensity. Now, we change the traffic intensity from a low to a high value which leads to a decrease of the PMRR $K(t)$ of the traffic aggregate. The simulation results are shown in Figure 4. Using the same two traffic types as before, we start with mean traffic shares $\alpha_1 = 0.2$ and $\alpha_2 = 0.8$ and swap them at simulation time $t_0 = 1000$ s to $\alpha_1 = 0.8$ and $\alpha_2 = 0.2$ by changing the type-specific request arrival rates. This increases the traffic intensity of the aggregate due to a change in the traffic mix $\alpha(t)$. In this simulation experiment, the QoS is at risk because flows with low traffic intensity are successively replaced by flows with high intensity and, therefore, the utilization of the link is increasing. Conventional EBAC (cf. Figure 4a) reacts again more slowly than EBAC with TSOB (cf. Figure 4b) although their

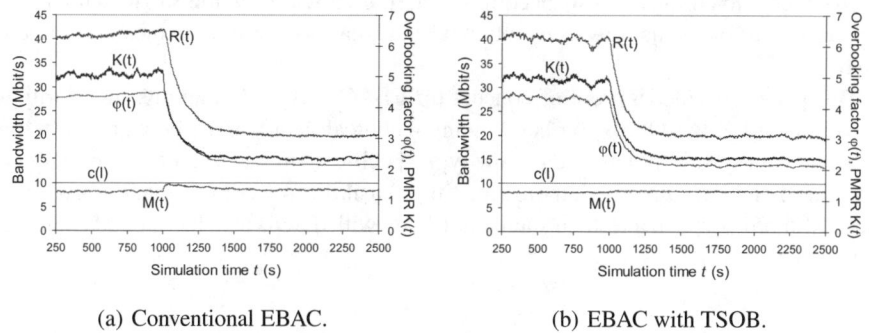

(a) Conventional EBAC.   (b) EBAC with TSOB.

Fig. 4. Conventional EBAC vs. EBAC with TSOB during a traffic intensity increase

response times differ less than in Figure 3. From other simulations [16] we know that the response time of conventional EBAC is independent of the EBAC memory in case of a sudden traffic increase. Our simluation results show that conventional EBAC yields a slightly higher link utilization compared to EBAC with TSOB. However, this high utilization comes at the expense of a violation of QoS guarantees as the measured traffic $M(t)$ consumes the entire link capacity $c(l)$ for a short period of time (cf. Figure 4a). As a consequence, the packet delay probability $p_d = P(\text{Packet delay} \geq 50 \text{ ms})$ rises from $p_d = 0$ for EBAC with TSOB to a maximum of $p_d \approx 0.3$ for conventional EBAC.

6   Conclusion

We reviewed the concept of experience-based admission control (EBAC) and summarized previous work regarding its robustness and adaptivity. EBAC overbooks the capacity of a single link with reservations according to the average peak-to-mean rate ratio of all admitted flows if the reservations are made based on signaled peak rates. The contribution of this paper is the extension of EBAC to use a compound type-specific overbooking factor for different traffic types subsuming flows with similar peak-to-mean rate ratios. The major challenge is the calculation of the type-specific reservation utilizations required for the compound overbooking factor. In general, the traffic cannot be measured type-specific and, as a consquence, the type-specific reservation utilizations cannot be obtained directly. Therefore, we proposed a least squares approximation to calculate the type-specific reservation utilizations depending on the reservation utilization of the entire traffic aggregate and the reserved rates of the type-specific aggregate shares. Our simulation results revealed that this method estimates with sufficiently high accuracy.

We simulated sudden and extreme changes of the traffic mix such that the share of flows with highly utilized reservations suddenly decreases or increases. If the share of these flows decreases, EBAC with type-specific overbooking (TSOB) adapts faster than conventional EBAC which leads to a significantly better resource utilization during the adaptation phase. If the share of these flows decreases, the advantage of EBAC with

TSOB over conventional EBAC becomes even more obvious: while EBAC with TSOB can avoid overload situations, conventional EBAC has no appropriate means to prevent them.

This paper provided a proof of concept for EBAC with TSOB and its superiority to conventional EBAC. On the on hand, many technical details must be clarified before it can be deployed in practice, e.g. how type-specific aggregates can be identified. On the other hand, we already demonstrated the feasibility of conventional EBAC by a successful prototype in a testbed such that EBAC with TSOB also has a good chance to be feasible.

Acknowledgements

The authors would like to thank Prof. Phuoc Tran-Gia for the stimulating environment which was a prerequisite for this work.

References

1. Menth, M.: Efficient Admission Control and Routing for Resilient Communication Networks. PhD thesis, University of Würzburg (2004)
2. Fidler, M., Sander, V.: A Parameter Based Admission Control for Differentiated Services Networks. Computer Networks **44**(4) (2004) 463–479
3. Boudec, J.L.: Application of Network Calculus to Guaranteed Service Networks. IEEE Transactions on Information Theory **44**(3) (1998)
4. Wroclawski, J.: RFC 2210: The Use of RSVP with IETF Integrated Services (1997)
5. Qiu, J., Knightly, E.: Measurement-Based Admission Control with Aggregate Traffic Envelopes. IEEE Transactions on Networking **9**(2) (2001) 199–210
6. Más, I., Karlsson, G.: PBAC: Probe-Based Admission Control. In: 2nd International Workshop on Quality of future Internet Services (QofIS). (2001)
7. Mandjes, M., van Uitert, M.: Transient Analysis of Traffic Generated by Bursty Sources, and its Application to Measurement-Based Admission Control. Telecommunication Systems **15**(3-4) (2000) 295–321
8. Breslau, L., Jamin, S., Shenker, S.: Comments on the Performance of Measurement-Based Admission Control Algorithms. In: IEEE Conference on Computer Communications (INFOCOM). (2000) 1233–1242
9. Cetinkaya, C., Knightly, E.: Egress Admission Control. In: IEEE Conference on Computer Communications (INFOCOM). (2000) 1471–1480
10. Elek, V., Karlsson, G., Rönngren, R.: Admission Control Based on End-to-End Measurements. In: IEEE Conference on Computer Communications (INFOCOM). (2000) 1233–1242
11. Grossglauser, M., Tse, D.: A Framework for Robust Measurement-Based Admission Control. IEEE Transactions on Networking **7**(3) (1999) 293–309
12. Jamin, S., Shenker, S., Danzig, P.: Comparison of Measurement-Based Call Admission Control Algorithms for Controlled-Load Service. In: IEEE Conference on Computer Communications (INFOCOM). (2000) 973–980
13. Gibbens, R., Kelly, F.: Measurement-Based Connection Admission Control. In: 15th International Teletraffic Congress (ITC), Washington D. C., USA (1997)

14. Milbrandt, J., Menth, M., Oechsner, S.: EBAC - A Simple Admission Control Mechanism. In: 12th IEEE International Conference on Network Protocols (ICNP), Berlin, Germany (2004)

15. Menth, M., Milbrandt, J., Oechsner, S.: Experience-Based Admission Control. In: 9th IEEE Symposium on Computers and Communications (ISCC), Alexandria, Egypt (2004)

16. Milbrandt, J., Menth, M., Junker, J.: Performance of Experience-Based Admission Control in the Presence of Traffic Changes. In: IFIP-TC6 Networking Conference (NETWORKING), Coimbra, Portugal (2006)

17. Roberts, J., Mocci, U., Virtamo, J.: Broadband Network Teletraffic - Final Report of Action COST 242. Springer, Berlin, Heidelberg (1996)

18. Law, A.M., Kelton, W.D.: Simulation Modeling and Analysis. McGraw-Hill, Boston, USA (2000)

19. Paxson, V., Floyd, S.: Wide-Area Traffic: The Failure of Poisson Modeling. IEEE/ACM Transactions on Networking 3(3) (1995) 226–244

20. Blake, S., Black, D., Carlson, M., Davies, E., Wang, Z., Weiss, W.: RFC2475: An Architecture for Differentiated Services (1998)

21. Rosen, E., Viswanathan, A., Callon, R.: RFC3031: Multiprotocol Label Switching Architecture (2001)

22. Bjorck, A.: Numerical Methods for Least Squares Problems. SIAM Society for Industrial & Applied Mathematics (1996)

23. Martin, R., Menth, M.: Improving the Timeliness of Rate Measurements. In: 12th GI/ITG Conference on Measuring, Modelling and Evaluation of Computer and Communication Systems (MMB) together with 3rd Polish-German Teletraffic Symposium (PGTS), Dresden, Germany (2004)

Applying Blood Glucose Homeostatic Model Towards Self-management of IP QoS Provisioned Networks

Sasitharan Balasubramaniam[1], Dmitri Botvich[1], William Donnelly[1], and Nazim Agoulmine[2]

[1] Telecommunications Software & Systems Group, Waterford Institute of Technology, Carriganore Campus, Waterford, Ireland
{sasib, dbotvich, wdonnelly}@tssg.org
[2] Networks and Multimedia Systems Group, University of Evry Val d'Essonne, Rue du Pelvoux, Evry Croucouronnes, France
Nazim.Agoulmine@iup.uni-evry.fr

Abstract. Due to the rapid growth of the Internet architecture and the complexities required for network management, the need for efficient resource management is a tremendous challenge. This paper presents a biologically inspired self-management technique for IP Quality of Service (QoS) provisioned network using the blood glucose regulation model of the human body. The human body has the capability to maintain overall blood glucose level depending on the intensity of activity performed and at the same time produce the required energy based on the fitness capacity of the body. We have applied these biological principles to resource management, which includes (i) the ability to manage resources based on predefined demand profile as well as unexpected and fluctuating traffic, and (ii) the ability to efficiently manage multiple traffic types on various paths to ensure maximum revenue is obtained. Simulation results have also been presented to help validate our biologically inspired self-management technique.

1 Introduction

Biological processes have tremendous capabilities to adapt to environmental changes due to their robust characteristics. This essential capability is largely due to fluctuating environment that living organisms must face in order to maintain survivability. Due to this reason a number of biological analogies have been applied towards communication networks (e.g. sensor networks, MANETS). One suitable application of biological analogies that provides an attractive solution towards supporting self-governance is autonomic network management. The current trend for network management requires autonomic capabilities that are exhibited through self-governance behaviour. One crucial requirement of self-governance is the ability for communication systems to self-manage and adapt to changes from the environment (e.g. changes in traffic demand or business goals).

In this paper we use principles of maintaining blood glucose in the human body as a mechanism to maintain network resource equilibrium. The human body maintains the blood glucose level irrespective of any behavioural changes (e.g. light activity to

G. Parr, D. Malone, and M. Ó Foghlú (Eds.): IPOM 2006, LNCS 4268, pp. 84–95, 2006.

non-routine heavy exercises). We compare this analogy to the intensity usage of networks, such as the ability to handle routine demand traffic and unexpected traffic requests. We view the intensity usage of network resources from the different traffic types and the quantity of each traffic type (e.g. small data and multimedia is low intensity compared to large data and multimedia stream). At the same time we also consider the revenue output of the different traffic intensity and compare this to the energy production from the body, which is essentially determined from the Anaerobic and Aerobic respiration of the body. By applying these analogies we can show how networks resource management can adjust dynamically to fluctuating traffic demands to support high level business goals of Internet Service Providers (ISP).

The paper is organised into the following subsections. Section 2 reviews current related work for resource management as well as bio-inspired analogies applied to communication networks. Section 3 briefly describes the blood glucose homeostasis while section 4 describes the application of this analogy towards network resource management. Section 5 describes results from our simulation to validate our idea. Finally section 6 presents the conclusion and future work.

2 Related Work

2.1 Bio-inspired Network Management

Employing biological analogies towards telecommunications networks has gained tremendous popularity in recent times. Suzuki and Suda [1] applied the analogy of bee colony behavior for the bio-networking architecture for autonomous applications, where network applications are implemented as a group of autonomous diverse objects called cyber-entity (CE). Leibnitz et al [2] proposed a biological inspired model to solve the problem of multi-path routing in overlay networks by adapting the transmission packets to changes in the metrics of each path. The solution proposed using multiple primary paths to route traffic and switching between the different primary paths depending on the load on each path. However, the solution is based purely on selecting paths based on their loads and not considering fluctuations. At the same time, the proposed solution assumes equal priority between different traffic types. A number of biologically inspired mechanisms have also been applied to route management in networks, especially analogies that mimic social insect behaviour. An example is the work called *AntNet* by Di Caro and Dorigo [3] which employs a set of mobile agents mimicking ant behaviour to probe routes and update routing tables.

2.2 IP Network Resource Management

Mantar et al [4] proposed a Bandwidth Broker (BB) model to support QoS across different DiffServ Domains. The architecture is based on a centralised controller and uses centralised network state maintenance and pipe-based intra-domain resource management scheme. However, their solution has only considered one QoS class and has not provided mechanisms to dynamically determine new paths once the original paths are congested. Gojmerac et al [5] proposed an adaptive multipath routing mechanism for dynamic traffic engineering. The solution is based on load balancing technique that uses local interactions between the devices, and disseminating congestion information through backpressure messages. Although the solution is based on decentralised control of resource management, the solution does not consider how diverting traffic due to congestion on particular links, could affect other paths which may handle traffic from the demand profile. Since, the back propagation message gets sent recursively through all the nodes, the time required to determine new paths is relatively slow and not reactive to fluctuating traffic. Yagan and Tham [6] proposed a self-optimising, self-healing architecture for QoS provisioning in Differentiated services. The architecture employs a model free Reinforcement Learning (RL) approach to counter the dimensionality problems of large state spaces found in conventional Dynamic Programming (DP) techniques. Simulation results of their work have shown that this solution is not suitable for dynamic networks.

3   Blood Glucose Homeostasis and Respiration

3.1   Blood Glucose Homeostasis

In this section we will describe key biological principles that we will apply as analogy towards self-management for autonomic networks. Organisms have the ability to maintain system equilibrium, which is also known as Homeostasis [7]. The process for balancing homeostasis is through positive and negative feedback loops, where the amount of resources that is balanced is dependent on the intensity of the activity performed by the human body. When the body is going through various activities, the blood glucose is balanced in the body by obtaining glucose from various sources once current glucose storage is depleted. This concept is illustrated in Fig. 1.

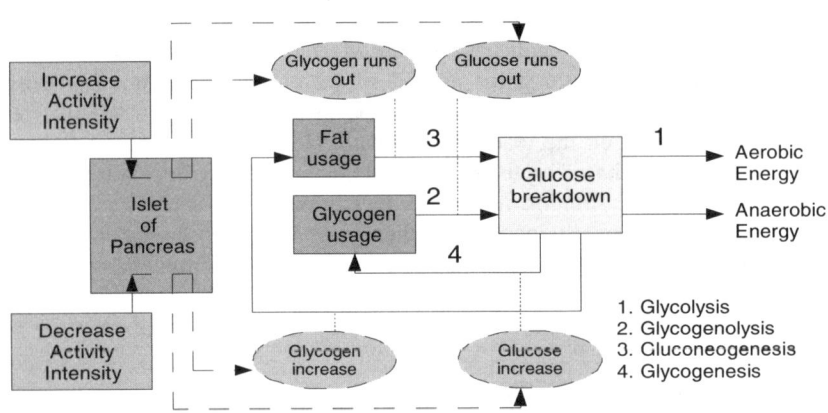

Fig. 1. Blood Glucose Homeostasis

These sources includes the liver, where glucose is obtained from glycogen or body fat. Glycogen is a storage form of glucose and is usually stored in the liver at large amount. Blood glucose is used to create energy through two respirations, which includes Aerobic and Anaerobic. As shown in Fig. 1, there are various chemical reactions used to maintain the blood glucose level. In the event that the intensity of body activity rises and need blood glucose to generate energy, the process of **Glycolysis** is performed. When Glucose in the blood runs out, the Glucose is obtained from Glycogen through the process of **Glycogenolysis**. The **Gluconeogenesis** is the generation of glucose from other organic molecules such as fat, and occurs once the glycogen is used beyond a specific threshold. We refer to the fat that is discovered and converted to glucose during Gluconeogenesis as "good fat". In the event that large amount of glucose is found in the blood, the blood glucose level is reduced by transforming to glycogen through **Glycogenesis,** which usually occurs when the intensity of the body activity decreases. In the event that the amount of glycogen increases beyond a particular threshold, this is transformed into fat. Once this fat goes beyond a specific threshold, this will amount to extra fat that can lead to an unhealthy state of the body which we refer to as "bad fat". Therefore, the glucose is obtained from various forms and used as energy depending on the intensity of the body, and in the event that the activity is reduced, this glucose is transformed back into various forms (e.g. glycogen or fat).

3.2 Respiration

Respiration is the process of creating energy by converting energy-rich molecules such as glucose into energy. There are predominantly two types of respiration, which are Aerobic and Anaerobic respiration. Aerobic respiration is more efficient than Anaerobic respiration and is the preferred method of glycogen breakdown. Aerobic respiration requires oxygen to generate energy. The resulting energy from Aerobic respiration is usually very high (2830 KJ mol-1), and is used to fuel long term high intensity body workout. Body that is usually regarded as fit will tend to have longer period of Aerobic respiration. The human metabolism process is primarily Aerobic, but during Anaerobic conditions the overworked muscles that are starved of oxygen creates very low energy, and usually occurs towards the end of maximum body intensity. The Anaerobic respiration usually leads to small amount of energy (118 KJ mol-1) compared to Aerobic respiration.

4 Network Model for Self-management

In this section we will map the biological principles described in section 3 to the self-management of resources in core networks. Our overall model consist of two layers, which includes (i) self-management of resources using blood glucose and effective management of multiple traffic types per paths using respiration analogies, and (ii) self-organisation for decentralised control using reaction-diffusion [8]. However, the focus of this paper will only concentrate on the resource management of the network as well as traffic class management per path.

The first principle of biological analogy mapping is to the overall management of resources of the entire network (this is compared to blood glucose homeostasis), and the second mapping is used to determine the optimum ratio of data and multimedia within links of each path (this is compared to respiration).

4.1 Comparison of Blood Glucose to Overall Resource Management

The same way that the human body self-manages blood glucose in the body depending on the activity and intensity use of the body, will be used as an analogy towards determining the mechanism that networks handle traffic at different intensity and its mechanism to manage spare resources. Our comparison to the blood glucose homeostasis feedback loop is shown in Fig. 2. We compare the usage of glucose to the normal operations of the network such as packet forwarding between nodes. The glycogen usage in the body is used once the glucose in the body is depleted; we compare this to the network using resources for load balancing and to support routine traffic from a demand profile. The demand profile contains the historical traffic statistics that is collected over a period of time and reflects the routine traffic between two edge routers (ER_i, ER_j). To support the demand profile a set of primary paths $(P_n, n = 1,..N)$, where N is the total number of primary paths required for the edge pairs, are formed for the pair of edge routers. When a traffic stream t_{ER_i, ER_j} is admitted between the edge pair (ER_i, ER_j), the traffic is routed along paths P_n.

In the event that the edge routers encounter unexpected traffic requests or fluctuations that are beyond the capacity of the primary paths P_n, the Resource Manager (RM) begins to discover new path along the spare capacity of the other links within the network. When we compare this process to the blood glucose homeostasis, this is similar to the body using up all its glycogen and must begin the fat (good fat) discovery process to obtain new source of glucose. The fat on the other hand is a source of glucose that can only be accessed in rare cases and will have to be discovered from other sources. In similar way, the discovery of spare capacity P_{SC,ER_i, ER_j} between the two edge pairs occurs in rare occasion.

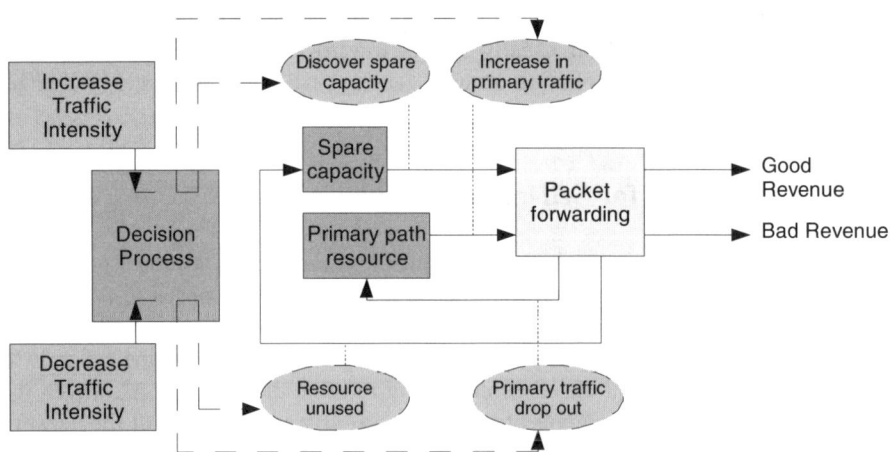

Fig. 2. Network resource self-management

An example of this process on a network is illustrated in Fig. 3. Initially we determine the amount of resource that we require with respect to a demand profile statistics. For example, if we have a demand profile between 6am to 6pm for data of 10Gb and multimedia of 10GB, then we may require total resource of 20Gb during the 12 hour period. As shown in Fig. 3, the demand profile may lead us to have two primary paths (P1 – 10Gb and P2 – 10Gb, which both gives 20Gb). Therefore, when traffic comes through the network, this resource is readily used up to support the requested traffic. When this is compared to the blood glucose model, the glycogen is broken down through **Glycogenolysis** and **Glycolysis** to support respiration and create energy. In the event that unexpected traffic comes through the network, we do spare capacity discovery (this is shown as path F1). This is similar to the human body when the body is pushed to the limit and the stored glycogen is depleted, the body discovers the fat and breaks this down to create glucose.

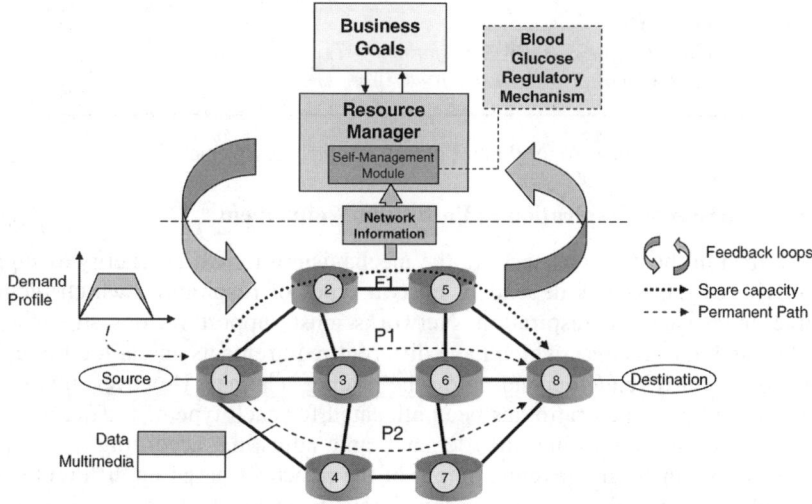

Fig. 3. Comparison of Primary streams to Spare capacity streams in network resource management

In the event that after a number of days the body tends to use the fat resource regularly, the fat is transformed into glycogen for long term use. This in turn reduces the amount of fat in the network, and is determined through a threshold T_{SPARE_USAGE}. The threshold determines the length of time the spare capacity resources is being used. This analogy fits well to the human body, where as the body is being exercised the good fat is reduced which leads to a fitter body. At the same time in the event that demand traffic starts to drop out and the 20Gb demand profile does not get used up to its full capacity, the left over resources will eventually lead to revenue loss. This is compared to the human body that has excess amount of resources leading to production of bad fat. The resulting revenue loss is an indication to the ISP of over subscription of resources. An algorithm to describe these mechanisms is shown in Fig. 4.

```
/*Determine Primary paths*/
for edge pair (ERᵢ, ERⱼ), determine total request bandwidth (BW_{ERi, ERj}) from Demand Profile
        Determine all possible primary paths Pₙ (n = 1,..N), for BW_{ERi, ERj} between edge pairs
        (ERᵢ, ERⱼ) using shortest path algorithm based on link weigh = 1/capacity

/*Routing of traffic request*/
for new traffic t_{i,ERi,ERj} of request i with bandwidth BW_{t,i}, route through Paths Pₙ
        add BW_{t,i} to used bandwidth BW_{u,ERi, ERj}

/*Discovery of spare capacity*/
if ( total used bandwidth BW_{u,ERi, ERj} >= BW_{ERi, ERj} )
        start spare capacity discovery for new path P_{SC, ERi, ERj} for (ERᵢ, ERⱼ)
        if ( time of traffic use for P_{sc,ERi, ERj} , t_{Psc} > threshold T_{SPARE_USAGE} )
                Add P_{SC, ERi, ERj} to Pₙ

/*Determining bad fat*/
if ( used bandwidth BW_{u,ERi, ERj} < BW_{ERi, ERj} for time > T_{BAD FAT} )
        decrease (BW_{ERi, ERj} - BW_{u,ERi, ERj}) from BW_{ERi, ERj}
```

Fig. 4. Algorithm for overall resource management

4.2 Comparison of Respiration to Path Ratio Refinement

As described in the previous section, the mechanism for creating energy to support the activity of the user is dependent on two types of respiration, which includes Aerobic and Anaerobic respiration. Networks must support various stream types (e.g. data and multimedia), which have different requirements and at the same time outputs different revenue depending on the pricing schemes. However, this is also dependent on how much ratio has been allocated for each type of traffic, which in turn depends on monitoring the demand and adjusting according to demand changes. We compare the revenue output of the networks based on different traffic types and their intensity, to the energy output from the body depending on the respiration capabilities. Fig. 5 provides an illustration of our proposed solution for path ratio refinement, where the cross section of each path and the ratio of data and multimedia per path is shown.

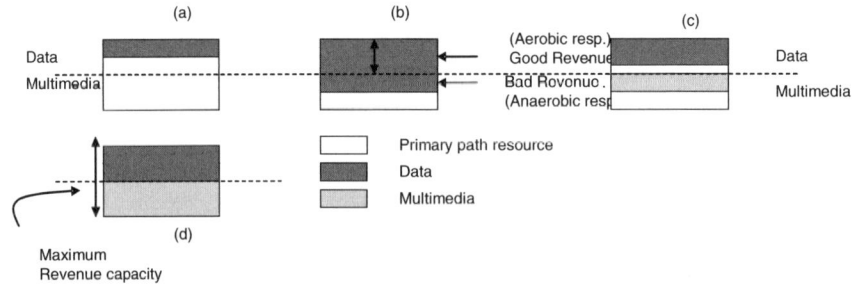

Fig. 5. The fitness test of path ratio refinement

The same way that we measure the fitness of the human body to generate energy depending on the activity, we compare this to the fitness of the paths allocated in the network by the ISP subscription for a pair of edge routers. Our definition when compared to the human body is the ability for the body to maximise Aerobic respiration to the point that oxygen runs out and switches to Anaerobic respiration. Therefore, we refer to the ability for specific allocated resources to support the same type of traffic as good revenue (R_{GOOD}), but when different type of traffic uses resource allocated to different traffic, this in turn gives bad revenue (R_{BAD}). We compare this to the oxygen supply to permit Aerobic respiration (Fig. 5b shows this comparison), where our aim is to maximise Aerobic respiration. An algorithm of our process is shown in Fig. 6.

```
/*Accept traffic stream and allocate right resource*/
for new traffic t request admission
        if ( traffic of type data t_{D,i,ER_i,ER_j} between edge routers (ER_i, ER_j) &&
            Data_{current_Resource} < Data_{Threshold} )
                Allocate into data buffer and transmit along link
                Update Good Revenue R_{Good} = R_{Good} + BWt_{D,i,ER_i,ER_j}
        else if ( traffic of type multimedia t_{M,i,ER_i,ER_j} between edge routers (ER_i, ER_j) &&
            Multimedia_{current_Resource} < Multimedia_{Threshold} )
                Allocate into multimedia buffer and transmit along link
                Update Good Revenue R_{Good} = R_{Good} + BWt_{M,i,ER_i,ER_j}

/*Allocating traffic stream to different resource*/
for new traffic request admission
        if (traffic of type data t_{D,i,ER_i,ER_j} between edge routers (ER_i, ER_j) &&

            Data_{current_Resource} > Data_{Threshold})
                Allocate into multimedia buffer and transmit along link
                Update Bad Revenue R_{BAD} = R_{BAD} + BWt_{D,i,ER_i,ER_j}
        else if ( traffic of type multimedia t_{M,i,ER_i,ER_j} between edge routers (ER_i, ER_j) &&
            Multimedia_{current_Resource} > Multimedia_{Threshold})
                Allocate into data buffer and transmit along link
                Update Bad Revenue R_{BAD} = R_{BAD} + BWt_{M,i,ER_i,ER_j}

/*Changing link ratio*/
for new traffic request admission
        if ( time at Bad Revenue, T_{Bad Revenue} > T_{refinement} )
                Calculate new ratio = BWt_{D,i,ER_i,ER_j}/BW_{Pn}
                Change scheduling based on new ratio
```

Fig. 6. Algorithm for path ratio fitness

An example of allocated resource being used by the corresponding traffic is shown in Fig. 3a, c, and d, where the amount of resource usages is at its optimum because the revenue output is at its optimum (e.g. only R_{Good} - all Aerobic and no Anaerobic respiration). In Fig. 3d, the maximum revenue is obtained when all the resources are being used for their allocated traffic type (e.g. $R_{MAX. Allocated} = R_{Good}$), indicating the maximum fitness of the path. When the network is at its maximum fitness, this indicates to the ISP providers that maximum revenue is being obtained. The ISP providers can determine the amount of fitness from historical readings to determine if they are receiving maximum Aerobic respiration (full fitness), slight Anaerobic and Aerobic respiration (slight unfitness), or usage of fat (unfit network that needs more resources to prolong fitness).

Fig. 3b shows an example of a particular type of traffic that has spilled over the allocated threshold, which means that the allocated resource is being used by a different traffic type. Fortunately, the multimedia resource allocated is currently under utilised allowing the extra data to use this resource. However, since the resource being used is of a different type, this has resulted in R_{BAD} (e.g. $R_{MAX.\ Allocated} \geq R_{BAD}$). If this behaviour continues, the Resource Manager will transform the ratio and possibly change charging schemes to maximise good revenue. This could lead to ISP changing the threshold of the allocated resources. In the event that this resource usage is short term, then there is no effect in long term changes.

5 Simulation Experiments

We have performed simulation work to validate our Bio-inspired resource management algorithms. The topology used in our simulation is shown in Fig. 7, and the routing paths are shown in Table 1. The simulation we have performed is to test the effectiveness of managing the resources within the network based on demand profile and the ability to handle fluctuations. The simulator follows our algorithm where initially from the demand profile we determine the possible routes from the different source and destination pairs. In our particular case, we have three pairs, which includes (S1 – D1), (S2 – D2), and (S3 – D3). We will concentrate on the performance of pair S1 – D1, where the other two pairs will be used to transmit background traffic in the network. The demand traffic that we inject into the network is shown in Fig. 8, while Fig. 9 shows the amount of resources utilized by each of the paths. Initially, the paths are pre-determined using our shortest path algorithm based on inverse bandwidth (link weight = 1/capacity), as described in the algorithm presented in the previous section.

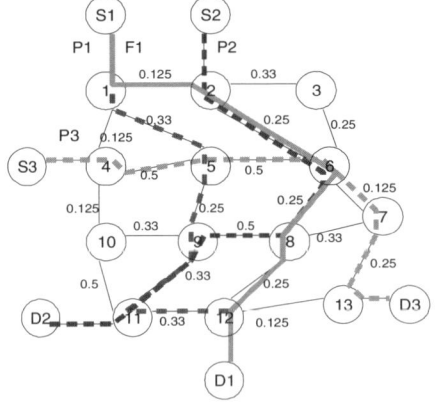

Table 1. Simulated Routing paths

Paths	Route
P1 (S1 – D1)	1 – 5 – 9 – 11 – 12
P2 (S2 – D2)	2 – 6 – 8 – 9 – 11
P3 (S3 – D3)	4 – 5 – 6 – 7 – 13
F1 (S1 – D1)	1 - 2 – 6 – 8 – 12

Fig. 7. Simulated Topology

Each of the path corresponds to a path for a pair of edge nodes. For example path P1, is routed through paths *1 – 5 – 9 – 11 – 12* with maximum bandwidth of *0.25* Mbps. As shown in Fig. 8, the streams that are transmitted through the network increases with time for all paths. At time 15, the amount of resource usage for path 1 exceeds the maximum capacity of 0.25Mbps (combination of data and multimedia stream), where at time 15 a new multimedia stream of 0.03 Mbps that is not part of the demand profile is injected into the network. At this point in time, the shortest path algorithm is executed to determine new spare capacity that is available between edge routers 1 and 12.

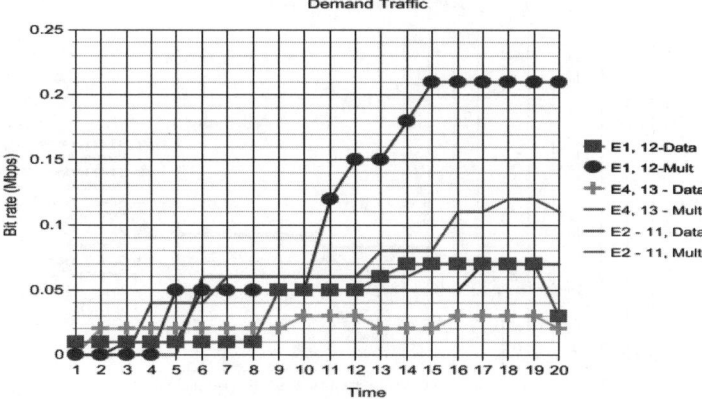

Fig. 8. Simulated demand traffic

Fig. 9 shows this process as the *S1-D1 Spare* line, which begins at time 15.

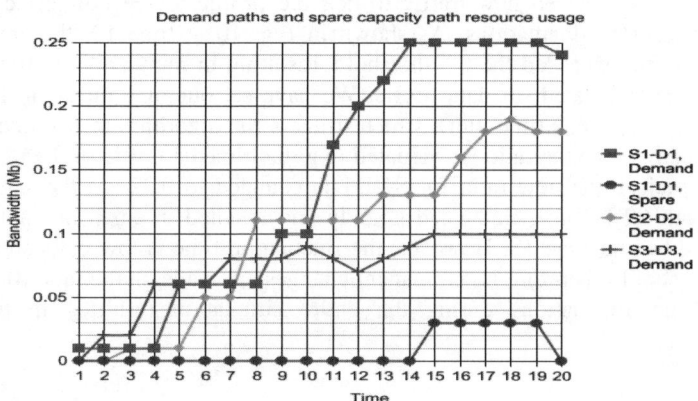

Fig. 9. Demand path and spare capacity path resource usage

As shown in table 1, this new path (F1) takes on the path *1 − 2 − 6 − 8 − 12*. The amount of time that the path F1 is alive supports the need for the ISP to purchase more resources to handle the new traffic demand. We have set the minimum threshold of spare capacity usage ($T_{SPARE\_USAGE} = 2$ time units) to determine how much fluctuation is permitted before this new spare capacity is added to new permanent path between S1-D1. As shown in Fig. 9, the paths between S2-D2 and S3-D3 do not have traffic requests that go beyond their capacity, and therefore does not require spare capacity discovery.

As described in the previous section, we also test the fitness of the paths to determine the effective ratio between data and multimedia traffic on each path. This is shown in Fig. 10, which illustrates the good and bad revenue generated from the different paths.

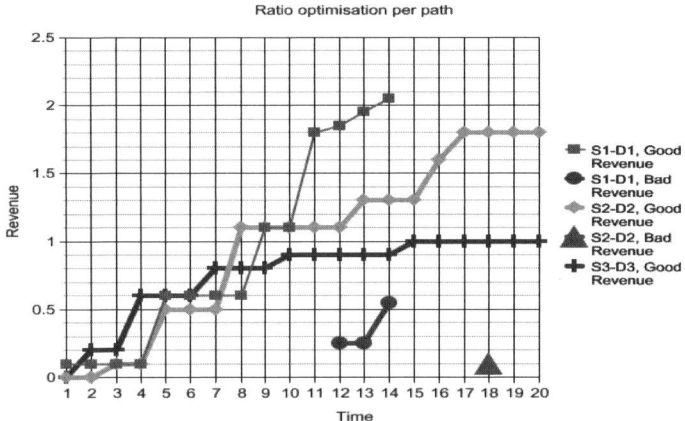

Fig. 10. Ratio optimization refinement for each path

Initially, we set the ratio for each path to 0.5 for both data and multimedia. During the simulation we injected new traffic to demand profile to see how effective our system will react to fluctuations. As shown in Fig. 10 at time 12, the amount of multimedia traffic exceeded the 0.5 threshold, resulting in multimedia traffic having to use resources allocated for data traffic. We have set our time threshold for ratio refinement ($T_{Refinement}$ = 2 time unit), which triggers the algorithm to re-calculate the ratio of the path. Once the ratio is evaluated (e.g. multimedia traffic of 0.18 Mb from 0.25 Mb capacity), the new ratio for S1-D1 is changed to 0.72. At the same time, traffic on path S2-D2 also fluctuated slightly and resulted in small amount of bad revenue as shown at time 18. However, the fluctuation time is low compared to the $T_{Refinement}$ threshold. Therefore no permanent changes are made on the ratio (0.5). Path S3-D3 had no fluctuations during the entire simulation resulting in no ratio refinement.

6 Conclusion and Future Work

Due to the immense complexities that are resulting from the accelerated growth of IP networks, efficient resource management is crucial towards maintaining overall stability. In this paper we have proposed mechanisms used for maintaining blood glucose as a technique towards maintaining overall stability in managing network resources for multiple traffic types. We have applied two biological principles towards our resource management scheme, which includes (i) management of permanent paths based on demand profiles and the ability to discover spare capacity for unexpected or fluctuating traffic, and (ii) management of multiple traffic classes on each path to support maximum revenue for ISP providers. The paper has also presented simulation results to demonstrate our idea for both points.

This paper has described preliminary results and findings for our autonomic network management program. Future work will include integrating blood glucose computational models and determining the effectiveness of the body's ability to manage resources as a comparison to test the effectiveness of our model used for network resource management. The future work will also include extending the current self-management mechanism to fully de-centralised control.

References

[1] J. Suzuki, T. Suda, "A Middleware Platform for a Biologically Inspired Network Architecture Supporting Autonomous and Adaptive Applications", IEEE Journal on Selected Areas in Communications, vol. 23, No. 2, February 2005.

[2] K. Leibnitz, N. Wakamiya, M. Murata, "Biologically inspired Self-Adaptive Multi-path routing in overlay networks", Communications of the ACM, vol. 49, no.3, March 2006.

[3] G. Di Caro, M. Dorigo, "AntNet: Distributed stigmergetic control for communication networks", Journal of Artificial Intelligence Research 9, 1998, pp 317-365.

[4] H. A. Mantar, J. Hwang, I. T. Okumus, S. Chapin, "A Scalable Intra-Domain Resource Management Scheme for Diffserv Networks", Proceedings of 2nd International Working Conference on Performance modeling and evaluation of heterogeneous networks, West Yorkshire, UK, July 2004.

[5] I. Gojmerac, T. Ziegler, F. Ricciato, P. Reichl, "Adaptive Multipath Routing for Dynamic Traffic Engineering", IEEE Globecom, San Francisco, USA, December 2003.

[6] D. Yagan, C.-K. Tham, "Self-Optimizing Architecture for QoS Provisioning in Differentiated Services", In Proceedings of the Second International Conference on Autonomic Computing (ICAC' 05), Seattle, WA, USA, June 2005.

[7] P. Raven, G. Johnson, J. Losos, S. Singer, "Biology", 6th edition, McGraw-Hill, 2002.

[8] S. Balasubramaniam, W. Donnelly, D. Botvich, N. Agoulmine, J. Strassner, "Towards integrating principles of Molecular Biology for Autonomic Network Management", Accepted for Hewlett Packard Open View University Association (HPOVUA) conference, Nice, France, May 2006.

New Mathematical Models for Token Bucket Based Meter/Markers*

Rafal Stankiewicz and Andrzej Jajszczyk

AGH University of Science and Technology, Kraków, Poland
{stankiewicz, jajszczyk}@kt.agh.edu.pl

Abstract. The paper presents analytical models for two types of token bucket based meter/markers used as building blocks of Assured Forwarding PHB in IP/DiffServ networks: srTCM and trTCM. The models enable quick finding of meter/marker characteristics under particular configuration and checking how parameter manipulations affect the characteristics of meter/markers. The models are validated by simulations with two types of traffic: TCP/ftp and web-like traffic.

1 Introduction

A meter/marker is one of Differentiated Services (DiffServ) [2] network elements used to make the realization of Assured Forwarding Per Hop Behavior (AF PHB) [8] possible.

In general, realizations of meter/markers can be classified as token bucket based or rate estimator based. The former uses one or more token buckets that are incremented periodically and decremented on packet arrival. The latter uses a dedicated mechanism to estimate the transient traffic rate and use the current estimation for marking decision. This paper deals with the following two token bucket based meter/markers: Single Rate Three Color Marker (srTCM) [10], Two Rate Three Color Marker (trTCM) [9]. They are the most representative and most commonly implemented token bucket based meter/markers.

Practical implementation of mechanisms supporting service quality assurance is not easy. Selection of particular mechanisms and configuration of network elements guaranteeing high network performance under various traffic conditions is not straightforward. In the case of DiffServ, many studies have shown that finding nodes' configuration resulting in a fair service differentiation is difficult. The above problems are, among other issues, the driving factors for analytical modeling of network elements. Analytical models help in better understanding of mechanisms and phenomena occurring in networks. They facilitate examination, planning and configuration of networks.

Less accurate models of srTCM and trTCM were presented in [18]. Validation with TCP/ftp traffic was presented there. This paper presents a novel and more accurate version of the models and enriches the experimental validation with simulations with web-like traffic.

---

* This work was partly supported by the European Union under the NoE EuroNGI Project FP6-507613.

G. Parr, D. Malone, and M. Ó Foghlú (Eds.): IPOM 2006, LNCS 4268, pp. 96–107, 2006.
© Springer-Verlag Berlin Heidelberg 2006

2   Basic Assumptions

The key assumption is that the time between arrivals of consecutive packets constituting the aggregate has the exponential distribution. Such an assumption enables the use of elements of the classical queuing theory in the mathematical model. Some researchers show that such an approach is justified in a heavily loaded network [3,4,19]. Cao et al. [3,4] showed that connection and packet arrival processes tend locally toward the Poisson distribution and time series of packet sizes and round trip times tend locally toward independence as the rate of new TCP connections increases. Moreover, a simplified and intuitive explanation can be given: in the congested link there are thousands of flows, therefore, packets belonging to a single flow are split up by packets from other flows and, what follows, correlation between consecutive packets in the aggregated stream is reduced. The traffic in a heavily loaded network is often smoothed and shows less bursty nature. Numerous researchers successfully used the Poisson model of packet arrivals in their work, e.g., [14,15,17]. There are also many arguments against Poisson modeling [16,22], so the model is developed and validated very carefully.

The distribution of a packet size in the Internet is bimodal or trimodal and often one peak is dominating [5,20]. For this reason, a constant packet size is often used in modeling [1,6,13,14]. The average packet size of about 500, 1000 or 1500 bytes is used by investigators. A similar approach is taken in this paper.

2.1   Single Rate Three Color Marker

The srTCM consists of two token buckets C and E with depths CBS and EBS, respectively. To adapt the classical queuing theory to modeling of srTCM easily, it was assumed that current states of the token buckets are expressed in the number of packets, rather than in bytes. The token count of bucket C is incremented by one CIR times per second. The token count is decremented by one on each packet arrival. Hence, the average rate B of emptying bucket is equal to the inverse of the average packet inter-arrival time and is expressed in packets per second. The token count of bucket E is incremented with a rate CIR as well but only if the bucket C is full. It is decremented with a rate B if the bucket C is empty.

Intuitively, a D/M/1/K queuing system would be used for modelling srTCM since the arrival rate is deterministic, i.e., it is a constant rate equal to CIR. However, the experimental results showed that srTCM characteristics based on D/M/1/K fit experimental results considerably worse than those based on M/M/1/K. Therefore, the latter model was chosen.

To find marking probabilities for srTCM a state transition graph for such a system was developed (Fig. 1). The system consists of two M/M/1/K queues. If a queue representing bucket C is in state i and queue representing bucket E is in state j then the system is in state $E_{i,j}$. If a steady state of the system is

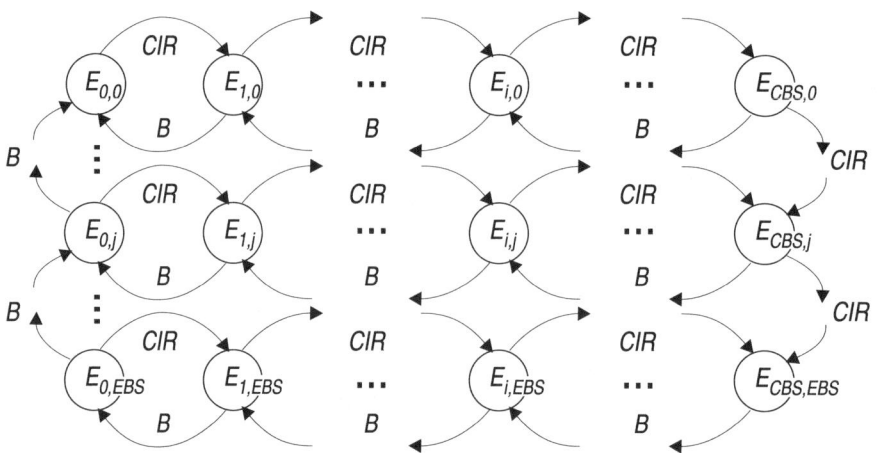

Fig. 1. State transition graph of the model of srTCM

assumed and $\pi_{i,j}$ denotes the probability that the system is in state $E_{i,j}$, the system can be described by the following set of algebraic equations:

$$0 = -CIR\,\pi_{0,0} + B\,\pi_{0,1} + B\,\pi_{1,0}$$
$$0 = -(CIR + B)\,\pi_{0,j} + B\,\pi_{0,j+1} + B\,\pi_{1,j} \qquad j = 1\ldots EBS-1$$
$$0 = -(CIR + B)\,\pi_{0,EBS} + B\,\pi_{1,EBS}$$
$$0 = -(CIR + B)\,\pi_{i,j} + B\,\pi_{i+1,j} + CIR\,\pi_{i-1,j} \qquad \begin{array}{l} i = 1\ldots CBS-1 \\ j = 0\ldots EBS \end{array} \qquad (1)$$
$$0 = -(CIR + B)\,\pi_{CBS,0} + CIR\,\pi_{CBS-1,0}$$
$$0 = -(CIR + B)\,\pi_{CBS,j} + CIR\,\pi_{CBS-1,j} + CIR\,\pi_{CBS,j-1} \quad j = 1\ldots EBS-1$$
$$0 = -B\,\pi_{CBS,EBS} + CIR\,\pi_{CBS,EBS-1} + CIR\,\pi_{CBS-1,EBS}$$

Obviously, a normalization equation is needed:

$$\sum_{i=0}^{CBS}\sum_{j=0}^{EBS}\pi_{i,j} = 1 \qquad (2)$$

Solving 1 and 2, the following formulas for $\pi_{i,j}$ can be obtained:

$$\pi_{i,0} = \frac{\left(\dfrac{CIR}{B}\right)^i \displaystyle\sum_{k=0}^{CBS-i}\left(\dfrac{CIR}{B}\right)^k}{\displaystyle\sum_{l=0}^{CBS+EBS}\left(\dfrac{CIR}{B}\right)^l \displaystyle\sum_{l=0}^{CBS}\left(\dfrac{CIR}{B}\right)^l} \qquad \text{for } i = 0\ldots CBS \qquad (3)$$

$$\pi_{i,j} = \frac{\left(\dfrac{CIR}{B}\right)^{j+CBS}}{\displaystyle\sum_{l=0}^{CBS+EBS}\left(\dfrac{CIR}{B}\right)^l \displaystyle\sum_{l=0}^{CBS}\left(\dfrac{CIR}{B}\right)^l} \qquad \begin{array}{l}\text{for } i = 0\ldots CBS \\ \text{and } j = 1\ldots EBS-1\end{array} \qquad (4)$$

$$\pi_{i,EBS} = \frac{\left(\dfrac{CIR}{B}\right)^{CBS+EBS} \displaystyle\sum_{k=0}^{i} \left(\dfrac{CIR}{B}\right)^{k}}{\displaystyle\sum_{l=0}^{CBS+EBS} \left(\dfrac{CIR}{B}\right)^{l} \displaystyle\sum_{l=0}^{CBS} \left(\dfrac{CIR}{B}\right)^{l}} \qquad \text{for } i = 0 \ldots CBS \qquad (5)$$

According to [10], a packet is marked red if upon its arrival both buckets are empty, that is, the system is in state $E_{0,0}$. Hence, the probability P^{mr} that a packet is marked red is equal to $\pi_{0,0}$. The packet is marked yellow if arrives at the moment when bucket C is empty and bucket E is not empty, i.e., the system is in state $E_{0,j}$ where $j = 1 \ldots EBS$. Consequently, the probability P^{my} that a packet is marked yellow equals $\sum_{j=1}^{EBS} \pi_{0,j}$. In the other cases the packet is marked green.

Finally, the probabilities that a packet will be marked red P^{mr}, yellow P^{my} or green P^{mg} are as follows:

$$P^{mr} = \frac{1}{\displaystyle\sum_{l=0}^{CBS+EBS} \left(\dfrac{CIR}{B}\right)^{l}} \qquad (6)$$

$$P^{my} = \frac{\displaystyle\sum_{j=1}^{EBS} \left(\dfrac{CIR}{B}\right)^{j+CBS}}{\displaystyle\sum_{l=0}^{CBS+EBS} \left(\dfrac{CIR}{B}\right)^{l} \displaystyle\sum_{l=0}^{CBS} \left(\dfrac{CIR}{B}\right)^{l}} \qquad (7)$$

$$P^{mg} = 1 - \frac{1}{\displaystyle\sum_{l=0}^{CBS} \left(\dfrac{CIR}{B}\right)^{l}} \qquad (8)$$

The above formulas express packet marking probabilities as a function of srTCM parameters, CIR, CBS and EBS, and traffic rate B.

2.2   Two Rate Three Color Marker

The trTCM consists of two token buckets C and P with depths CBS and PBS and token counts T_C and T_P, respectively. Again, it was assumed that the current state of buckets is expressed in the number of packets. The token count of bucket C is incremented by one CIR times per second while the token count of bucket P is incremented PIR times per second. The average rate of emptying token bucket P is equal to the average rate of incoming traffic B.

In the case of trTCM, the choice between M/M/1/K and D/M/1/K is ambiguous. As presented in Section 3 the results depend on the type of traffic. Therefore, both models are described.

In the case of M/M/1/K queuing system, the probability of bucket P being in state j can be expressed by a well known formula:

$$\pi_j^P = \frac{\left(\dfrac{CIR}{B}\right)^j}{\displaystyle\sum_{l=0}^{PBS} \left(\dfrac{CIR}{B}\right)^l} \tag{9}$$

Bucket C is emptied only if bucket P is not empty, thus, the conditional probability that bucket C is in state i is as follows:

$$\pi_{i|T_P \neq 0}^C = \frac{\left(\dfrac{CIR}{B\left(1-\pi_0^P\right)}\right)^i}{\displaystyle\sum_{l=0}^{CBS} \left(\dfrac{CIR}{B\left(1-\pi_0^P\right)}\right)^l} \tag{10}$$

where π_0^P is the probability that token bucket P is in state 0 (is empty).

A packet is marked red if upon its arrival bucket P is empty. If bucket P is not empty and bucket C is empty then the packet is marked yellow. A packet is marked green in all other cases [9]. Hence, the probabilities that a packet will be marked red, yellow or green are as follows:

$$P^{mr} = \pi_0^P = \frac{1}{\displaystyle\sum_{i=0}^{PBS} \left(\dfrac{PIR}{B}\right)^i} \tag{11}$$

$$P^{my} = \left(1 - \pi_0^P\right) \pi_{0|T_P \neq 0}^C = \frac{1 - P^{mr}}{\displaystyle\sum_{i=0}^{CBS} \left(\dfrac{CIR}{B\left(1-P^{mr}\right)}\right)^i} \tag{12}$$

$$P^{mg} = 1 - P^{mr} - P^{my} = \left(1 - P^{mr}\right)\left(1 - \frac{1}{\displaystyle\sum_{i=0}^{CBS} \left(\dfrac{CIR}{B\left(1-P^{mr}\right)}\right)^i}\right) \tag{13}$$

State probabilities for the D/M/1/K system and, what follows, marking probabilities are more difficult to obtain. In fact, closed form formulas for state probabilities are very complex. So the way to find a solution is provided in steps.

State probabilities for the D/M/1/K system can be obtained from state probabilities for M/D/1/K system by using the property of symmetry between those systems [12]. Generally, the probability that D/M/1/K queue is in state i is equal to the probability that M/D/1/K queue is in state $K - i$:

$$\pi_i^{D/M/1/K}(\rho) = \pi_{K-i}^{M/D/1/K}(1/\rho) \tag{14}$$

where ρ is the system load. In turn, the state probabilities of the finite M/D/1 queue can be obtained form the state probabilities for the infinite M/D/1 system as follows [11,13]:

$$\pi_j^{M/D/1/K}(\rho) = \frac{\pi_j^{M/D/1/\infty}(\rho)}{\pi_0^{M/D/1/\infty}(\rho) + \rho G(K)}, \qquad j = 0 \dots K - 1$$

$$\pi_K^{M/D/1/K}(\rho) = 1 - \frac{G(K)}{\pi_0^{M/D/1/\infty}(\rho) + \rho G(K)} \tag{15}$$

where $G(K) = \sum_{j=0}^{K-1} \pi_j^{M/D/1/\infty}(\rho)$. Finally, the state probabilities in the infinite M/D/1 system can be calculated from the following formulas [7,11]:

$$\pi_0^{M/D/1/\infty}(\rho) = 1 - \rho,$$
$$\pi_1^{M/D/1/\infty}(\rho) = (1-\rho)(e^\rho - 1),$$
$$\pi_i^{M/D/1/\infty}(\rho) = (1-\rho)\sum_{j=1}^{i}(-1)^{i-j} e^{j\rho}\left(\frac{(j\rho)^{i-j}}{(i-j)!} + \frac{(j\rho)^{i-j-1}}{(i-j-1)!}\right), \quad i \geq 2 \tag{16}$$

Formula (16) is valid only for the system load $\rho < 1$. Nevertheless, formula (15) can be used for any ρ since the factor $1 - \rho$ appearing in formulas for $\pi_i^{M/D/1/\infty}$ gets reduced. The more formal way to obtain formula (15) is presented in [11].

Finally, the packet marking probabilities for trTCM modeled using D/M/1/K queue models can be calculated as follows:

$$P^{mr} = \pi_0^{D/M/1/PBS}\left(\frac{PIR}{B}\right) \tag{17}$$

$$P^{my} = (1 - P^{mr})\,\pi_0^{D/M/1/CBS}\left(\frac{PIR}{B(1-P^{mr})}\right) \tag{18}$$

$$P^{mg} = 1 - P^{mr} - P^{my} \tag{19}$$

3   Validation of the Models

The simulations were performed with ns-2 simulator version 2.27. The final processing of the simulation data as well as solving the model equations was performed with the MATHEMATICA package.

The steady-state type simulation was used. *The method of batch means* [21] of collecting data was used. The simulation warm-up period was respected. The batch size was chosen sufficiently high to avoid autocorrelations. Confidence intervals were calculated for each estimated parameter on confidence level 95%. Confidence intervals were ommited since they were very small and difficult to show in the scale of the figure.

Two types of traffic were used in the experiments. One set of simulations was performed with TCP/ftp-like traffic, i.e., traffic was generated by long-lived TCP sources, that is, always having unlimited data to send. A separate group of simulations were done with sources generating a web-like traffic. In both cases TCP Reno sources were used.

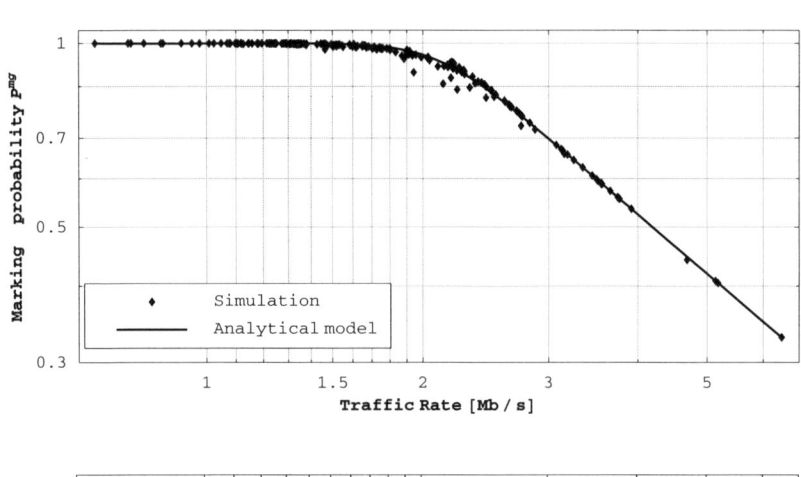

(a)

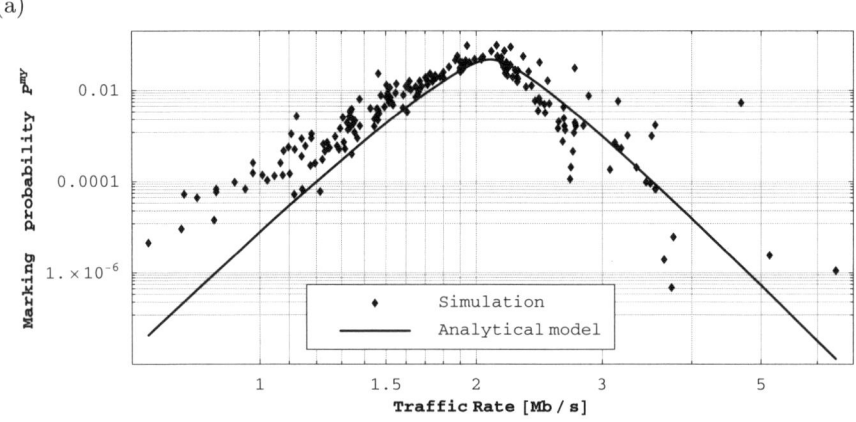

(b)

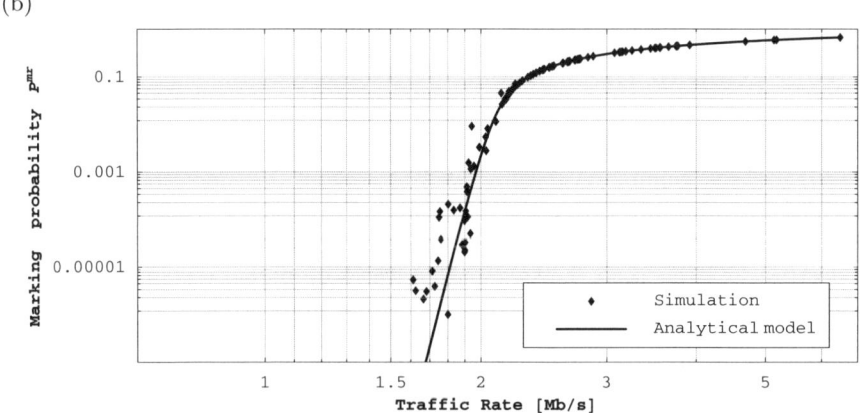

(c)

Fig. 2. Experiment 1: probability of marking packet as green (a), yellow (b) and red (c), TCP/ftp traffic

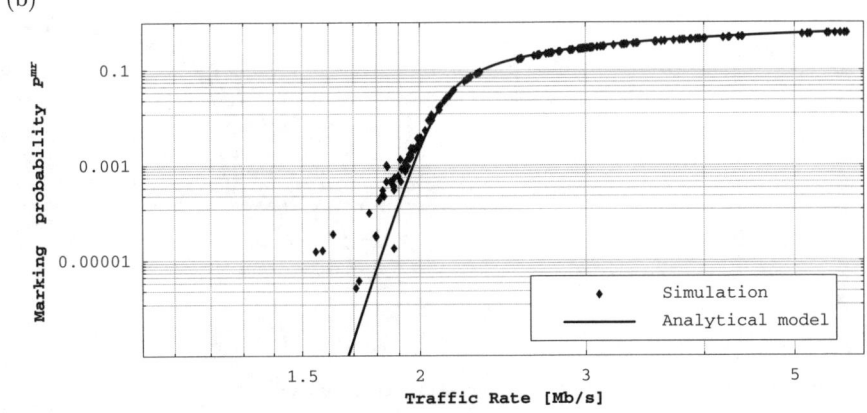

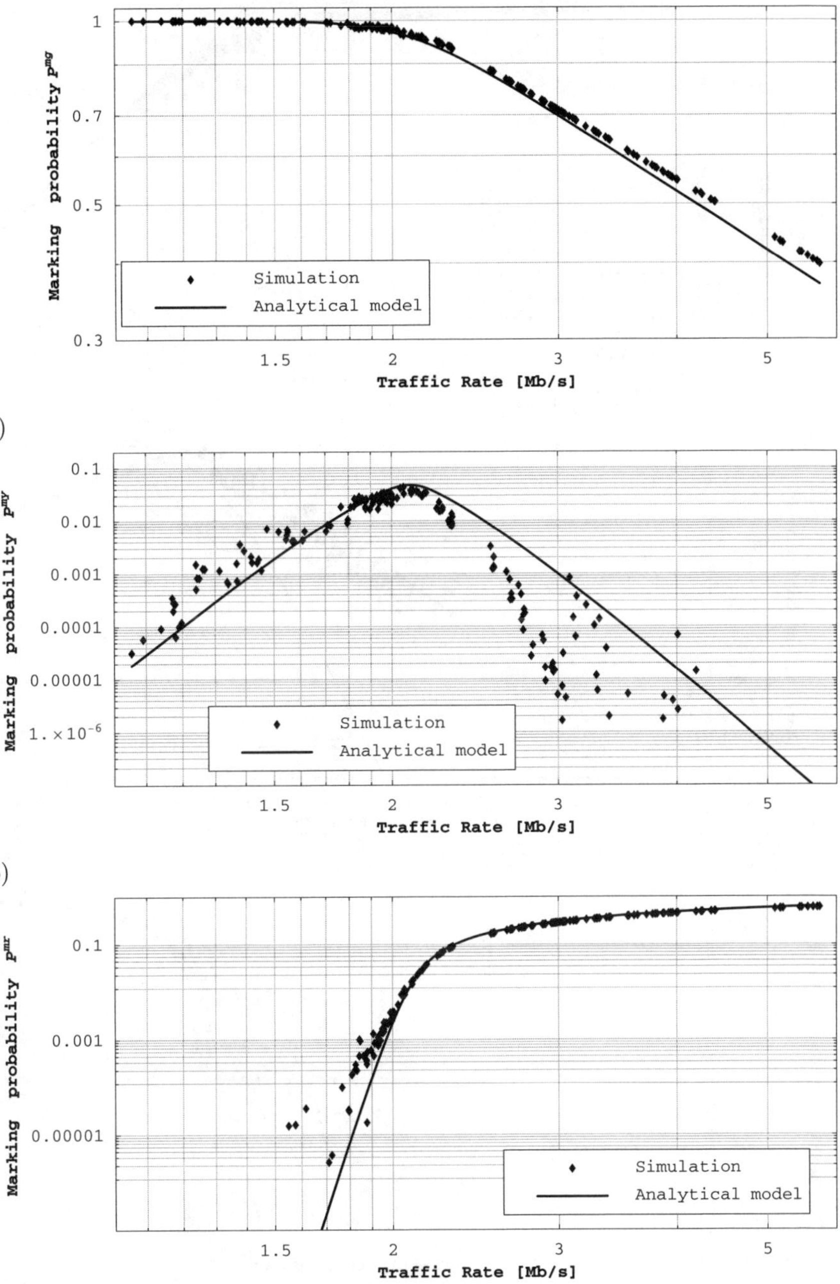

Fig. 3. Experiment 1: probability of marking packet as green (a), yellow (b) and red (c), web traffic

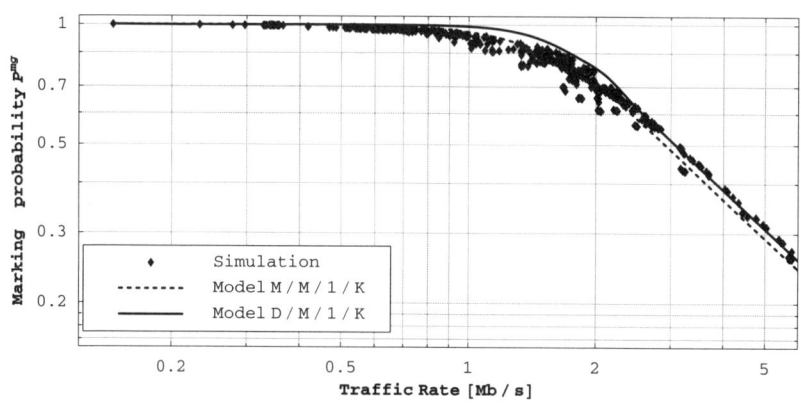

(a)

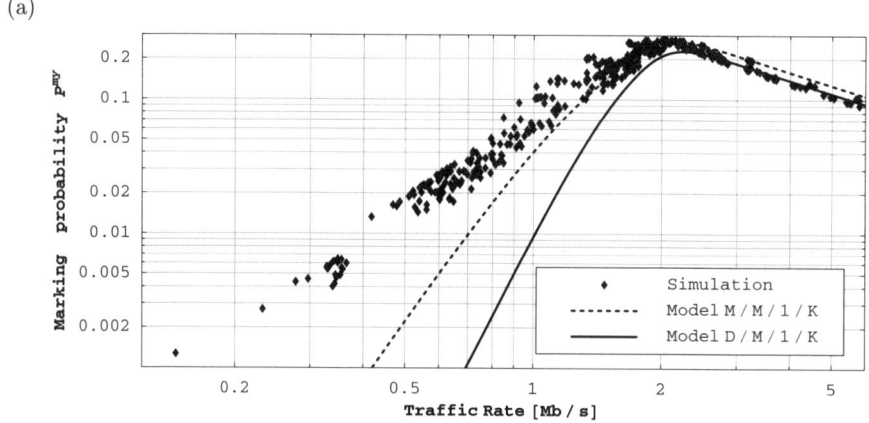

(b)

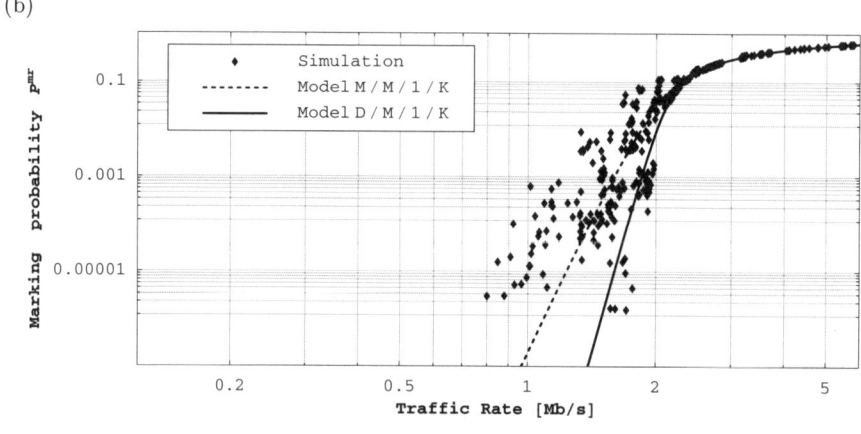

(c)

Fig. 4. Experiment 2: probability of marking packet as green (a), yellow (b) and red (c), TCP/ftp traffic

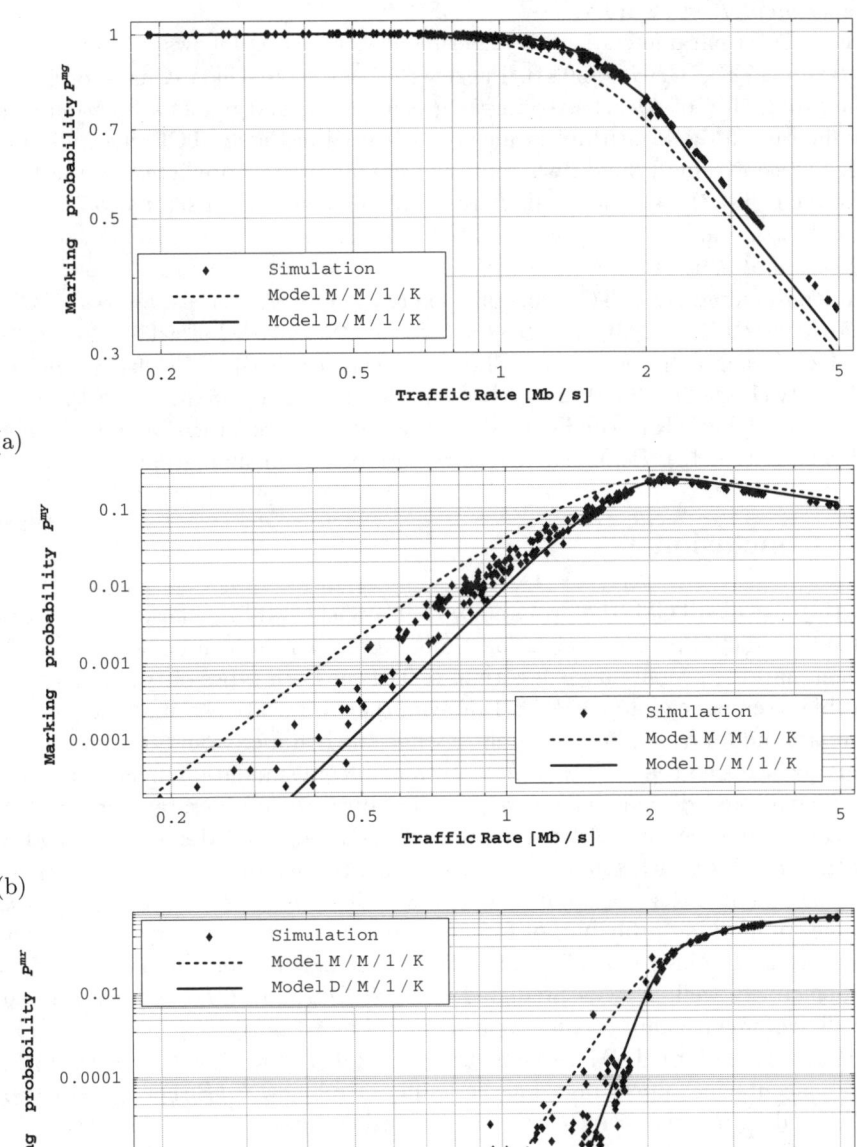

Fig. 5. Experiment 2: probability of marking packet as green (a), yellow (b) and red (c), web traffic

Experiment 1: srTCM

The srTCM parameters in this experiment were set as follows: $CIR = 2\,\mathrm{Mb/s}$, $CBS = 15\,\mathrm{kB}$, $EBS = 50\,\mathrm{kB}$. Packet size was set to $1\,\mathrm{kB}$. Characteristics of the probability of packet marking as green, yellow and red as a function of the traffic rate in the logarithmic scale are presented in Fig. 2 (TCP/ftp traffic) and Fig. 3 (web traffic). Solid lines represent characteristics predicted by the model (formulas 6 – 8) while the points are results obtained from simulation.

Experiment 2: trTCM

The parameters of trTCM in this experiment were set as follows $CIR = 1.5\,\mathrm{Mb/s}$, $PIR = 2\,\mathrm{Mb/s}$, $CBS = 5\,\mathrm{kB}$, $PBS = 20\,\mathrm{kB}$. Packet size was set to $1\,\mathrm{kB}$. Characteristics for trTCM are shown in Fig. 4 (TCP/ftp traffic) and Fig. 5 (web traffic). Solid lines represent characteristics predicted by the model based on D/M/1/K (formulas 17 – 19). Dashed lines are for the M/M/1/K model (formulas 11 – 13). Points show results obtained from simulation.

4   Conclusion

In the case of srTCM, the probability that a packet will be marked red or green is well predicted by the model. Some more discrepancies can be noticed for P^{mr} when marking probability is less than ≈ 0.01 for both types of traffic. The most visible discrepancies are for P^{my}. However, they are relatively small for the traffic rate around CIR and become greater while the distance from CIR increases. It must be stressed that values of P^{my} obtained from simulation in those areas are very small and do not make a regular line but are considerably spread. Thus, providing an exact formula describing P^{my} is not possible. Moreover, inexactness of the model for such small values of marking probability is not very critical. It can be assumed that for traffic rate $< 0.8\,CIR$ values of P^{mr} and P^{my} are ≈ 0. Values of marking probabilities for traffic rate about CIR and greater are more important. In that area P^{mr} and P^{mg} have higher values and are predicted satisfactorily well. P^{my} is predicted quite well around CIR but for a greater traffic volume it again becomes spread but close to zero.

Conclusions for trTCM are similar regarding P^{mr} and P^{mg}. In the case of P^{my} the inaccuracies appear only for the traffic rate below CIR. For rates between CIR and PIR and above PIR the model predicts characteristic of P^{my} well. The second difference is that the meter/marker behaves slightly differently for the two types of traffic. Clearly, the M/M/1/K model better fits simulation results for the TCP/ftp traffic while the D/M/1/K model fits results for the web-like traffic.

For the web-like traffic P^{mg} becomes increasingly underestimated if the traffic rate increases above CIR for both srTCM and trTCM but this discrepancy is not high.

The conclusion about meter/markers itself stemming from this observation is that they do not offer a strictly predictable marking probability if the marking probability is small.

References

1. A. A. Abouzeid and S. Roy. Modeling random early detection in a differentiated services network. *Computer Networks*, 40(4):537–556, Nov. 2002.
2. S. Blake, D. Black, M. Carlson, E. Davies, Z. Wang, and W. Weiss. An architecture for differentiated services. *IETF RFC 2475*, Dec. 1998.
3. J. Cao, W. S. Cleveland, D. Lin, and D. X. Sun. The effect of statistical multiplexing on internet packet traffic: Theory and empirical study. Technical report, Bell Labs, 2001.
4. J. Cao, W. S. Cleveland, D. Lin, and D. X. Sun. On the nonstationarity of internet traffic. In *Proc. ACM SIGMETRICS 2001*, pages 102–112, Jun. 2001.
5. R. Cáceres, P. Danzig, S. Jamin, and D. Mitzel. Characteristics of wide-area TCP/IP conversations. In *Proc. ACM SIGCOMM 1991*, pages 101–112, Sep. 1991.
6. S. Floyd and K. Fall. Promoting the use of end-to-end congestion control in the internet. *IEEE/ACM Transactions on Networking*, 7(4):458–472, Aug. 1999.
7. D. Gross and C. M. Harris. *Fundamentals of Queueing Theory*. John Wiley & Sons. Ltd, USA, 3rd edition, 1998.
8. J. Heinanen, F. Baker, W. Weiss, and J. Wroclawski. Assured forwarding PHB group. *IETF RFC 2597*, Jun. 1999.
9. J. Heinanen, T. Finland, and R. Guerin. A two rate three color marker. *IETF RFC2698*, Sep. 1999.
10. J. Heinanen and R. Guerin. A single rate three color marker. *IETF RFC2697*, Sep. 1999.
11. V. B. Iversen. *Teletraffic Engineering Handbook*. ITC/ITU-D SG 2/16, Sep. 2002.
12. J. Keilson. The ergodic queue length distribution for queueing systems with finite capacity. *Journal of the Royal Statistical Society*, 28(1):190–201, 1966.
13. P. Kuusela, P. Lassila, J. Virtamo, and P. Key. Modeling RED with idealized TCP sources. In *Proc. 9th IFIP Conference on Performance Modelling and Evaluation of ATM & IP Networks*, pages 155–166, Jun. 2001.
14. N. Malouch and Z. Liu. Performance analysis of TCP with RIO routers. In *Proc. IEEE GLOBECOM 2002*, pages 1633–1637, Nov. 2002.
15. M. May, J. Bolot, A. Jean-Marie, and C. Diot. Simple performance models of differentiated services schemes for the internet. In *Proc IEEE INFOCOM 1999*, pages 1389–1394, Mar. 1999.
16. V. Paxson and S. Floyd. Wide-area traffic: The failure of poisson modeling. *IEEE/ACM Transactions on Networking*, 3(3):226–244, Jun. 1995.
17. S. Sahu, D. Towsley, and J. Kurose. A quantitative study of differentiated services for the internet. In *Proc. IEEE GLOBECOM 1999*, pages 1808–1817, Dec. 1999.
18. R. Stankiewicz and A. Jajszczyk. Analytical models for DiffServ meter/markers. In *Proc. IEEE ICC 2005*, May 2005.
19. A. Stepanenko, C. C. Constantinou, T. N. Arvanitis, and K. Baughan. Statistical properties of core network internet traffic. *IEE Electronics Letters*, 38(7):350–351, Mar. 2002.
20. K. Thompson, G. Miller, and R. Wilder. Wide area internet traffic patterns and characteristics. *IEEE Network*, 11(6):10–23, Nov.-Dec. 1997.
21. J. Tyszer. *Object-Oriented Computer Simulation of Discrete-Event Systems*. Kluwer Academic Publishers, USA, 1999.
22. W. Willinger, M. S. Taqqu, R. Sherman, and D. V. Wilson. Self similarity through high-variability: Statistical analysis of ethernet lan traffic at the source level. *IEEE/ACM Transactions on Networking*, 5(1):71–86, Feb. 1997.

Unique Subnet Auto-configuration in IPv6 Networks

Reha Oguz Altug and Cuneyt Akinlar

Computer Engineering Department, Anadolu University, Eskisehir, Turkey
Phone: +90-222-321-3550x6553; Fax: +90-222-323-9501
{roaltug, cakinlar}@anadolu.edu.tr

Abstract. IPv6 host auto-configuration has been part of IPv6 specification from the start, but IPv6 routers still require manual configuration and administration. This is not only unacceptable for emerging
home and SOHO networks, but it also complicates network management for complex corporate networks. To enable easy and ubiquitous
deployment of future IPv6 networks, there is a need for an IPv6 router
auto-configuration protocol to complement IPv6 host auto-configuration
to make IPv6 networks truly plug-and-play. In this paper we address this
issue and propose an IPv6 router auto-configuration algorithm. The idea
is for each router to assign unique subnetids to each of their interfaces
during startup, and cooperate with other routers in the network to maintain the uniqueness of their subnetid assignment in the face of topological
changes. We show how the proposed algorithm can be implemented by a
simple modification of the basic intra-domain topology broadcast algorithm. Finally we extend the algorithm to hybrid networks, where some
of the segments are manually configured by the administrators and the
rest of the segments are auto-configured by the routers.

Keywords: IPv6, Zeroconf Networks, Address Auto-configuration.

1 Introduction

While IPv6 host auto-configuration is part of IPv6 specification by design, IPv6
router setup requires manual configuration and administration [1]. Although
there are proposals to automatically obtain the global IPv6 prefix for an organization, i.e., a site, from the Internet Service Provider (ISP) [8,9,10], IPv6 subnet
setup within the organization is still left to the administrators. As it stands,
administrators must manually configure unique IPv6 subnets over each segment
or link in the network, which would quickly get out of hand as the network scales
up. Worse yet, if the topology of the network changes, the entire network may
need to be reconfigured. Also for some emerging IPv6 networks such as home and
SOHO networks [4], there may not be any administrators, and the users may not
have the necessary technical skills to configure the network. The routers within
such networks must auto-configure to enable plug-and-play IPv6 networking.

We can classify IPv6 networks into two types: single-router networks and
multi-router networks. A single-router network is one where a router connects

G. Parr, D. Malone, and M. Ó Foghlú (Eds.): IPOM 2006, LNCS 4268, pp. 108–119, 2006.

several segments into a star topology, and also provides Internet connectivity, e.g., a simple home network. A multi-router network is one where two or more routers connect several segments into an arbitrary topology. One or more of the routers also provide Internet connectivity.

Although auto-configuration of a single-router IPv6 network is easy, auto-configuration of a complex multi-router IPv6 network is a formidable task. Current requirement for manual configuration and administration of such networks will stand in the way of ubiquitous deployment of IPv6 networks. Given that even a simple home or SOHO network may contain several routers due to multiplicity of link-layer technologies available at home and office, it would be desirable to have IPv6 router auto-configuration algorithms to automatically configure the entire IPv6 network without any manual intervention and administration.

In this paper we address IPv6 address auto-configuration of single and multi-router IPv6 networks with unique IPv6 subnets. The idea is to have each router randomly assign a unique IPv6 subnetid to each of their interfaces during startup, and maintain the uniqueness of their subnetid assignment in cooperation with other routers in the network. Due to local subnetid assignment, it is possible for two or more routers to assign the same IPv6 subnetid to different segments of the network thereby creating IPv6 subnetid conflicts. We show how routers can detect and resolve such IPv6 subnetid conflicts by augmenting an existing intra-domain routing algorithm, specifically the topology broadcast algorithm in section 5. Then in section 6 we show how the proposed algorithm can be modified to have both administered and auto-configured segments to co-exist in the same IPv6 network. This would be useful especially in complex IPv6 networks, where administrators may want to administer some of the segments and assign them manually-assigned IPv6 subnetids, and leave the rest of the segments to be auto-configured by the routers. Proposed auto-configuration algorithms would greatly simplify the management and deployment of such hybrid, complex IPv6 network topologies. We believe that the proposed algorithms will complement existing IPv6 host auto-configuration algorithm to enable truly plug-and-play IPv6 networking.

2 Unicast IPv6 Addresses and Terminology

An IPv6 address is 128-bits long, and is represented in what is called a colon-hexadecimal notation [1]. That is, each 16-bit block of an IPv6 address is converted into a 4-digit hexadecimal number and separated by colons. Figure 1 depicts the structures of 3 types of unicast IPv6 addresses [5,6,7].

A global unicast address consists of a 48-bit global routing prefix, a 16-bit subnetid and a 64-bit interface-id. An organization is assigned the global routing prefix; provided by an ISP during initialization using DHCPv6 [8,9], ICMPv6 [10] or some other mechanism. The subnetid is used within a site to identify IPv6 subnets. The subnetid configuration is left to the site's administrators and is currently done manually. The interface-id indicates the interface on a specific subnet within the site [5].

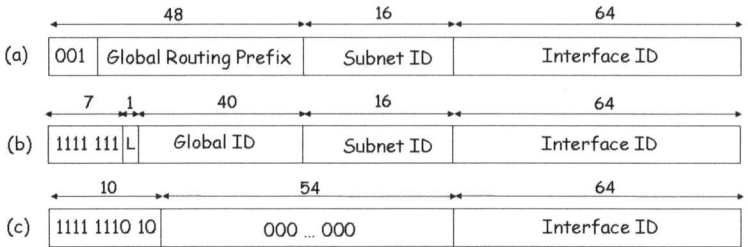

Fig. 1. Unicast IPv6 address structures: (a) Provider-based globally routable address, (b) Local unicast address, (c) Link-local address

As a replacement for site-local addresses [5], local unicast addresses have recently been standardized [7]. They are to be used if the site is disconnected from the Internet (thus lacking an ISP-based global prefix), or if sites wish to have a distinct prefix that can be used to localize traffic inside the site [7]. A local unicast address has a very similar structure to global routable address: The first 7 bits are fixed FC00::/7, followed by a bit named L which is set to 1, then followed by a 40-bit global ID. The last 64-bits are the interface-id. In the projected usage scenario, an administrator would randomly assign a 40-bit global ID to the site using the algorithm in [7], which is expected to be globally unique with a very small chance of collision. 16-bit subnetid would then identify the specific IPv6 subnet within the site as in global unicast addresses. It is expected that local unicast addresses and ISP-based global unicast addresses would share the same 16-bit subnetids, if they were being used concurrently [7]. Local unicast addresses are expected to be globally unique, but are intended to be used and routable within a site.

In addition to global and local unicast IPv6 addresses, each host must have a link-local IPv6 address. This is especially necessary for communicating between hosts on the same link with no router. A link-local address consists of a fixed 64-bit subnetid, i.e., FE80:/64, and a 64-bit interface-id. Since the subnetid is fixed, the address is auto-configured using the stateless address configuration mechanism described in section 3.

In the rest of this paper, we would denote a global or local unicast address as "p.s.i", where p is the first 48-bit prefix, s is the 16-bit subnetid, and i is the 64-bit interface-id.

3   IPv6 Host Auto-configuration Process

Host auto-configuration process is described in RFC 2462 [2]. An IPv6 host auto-configures its addresses either through (1) **stateless** address configuration, which is solely based on the receipt of Router Advertisements with one or more Prefix Information options, (2) **DHCPv6**, which is used when the Managed Address Configuration flag within a Router Advertisement is set to 1, or (3) using both of the mechanisms. Here are the steps of the IPv6 host auto-configuration process:

1. A tentative link-local address is derived based on the link-local prefix FE80::/64 and the 64-bit interface-id, and its uniqueness is tested over the link using Neighbor Solicitation messages.

2. After the unique link-local IPv6 address setup is complete, the host sends several (default three) Router Solicitation messages.

3. If no Router Advertisement messages are received, then the host uses DHCPv6 to obtain an address.

4. If a Router Advertisement message is received then

> **4.1.** For each Prefix Information Option with Autonomous flag set to 1, the host derives a tentative address using the prefix and the 64-bit interface-id, and tests its uniqueness. Once the uniqueness is verified, the host starts using the address with Preferred and Valid Lifetimes specified in the Router Advertisement.
>
> **4.2.** If Managed Address Configuration flag is set to 1, the host uses DHCPv6 to obtain an address.

As seen from above, after the link-local IPv6 address auto-configuration, the rest of the IPv6 host auto-configuration depends on Router Advertisements containing Prefix Information, which corresponds to the first 64-bits of an IPv6 address and specifies the IPv6 subnet for the segment. Recall from section 2 that 16-bit subnetid portion of an IPv6 prefix requires manual configuration by the administrators. In a complex network and especially dynamic IPv6 network, this could be very cumbersome. It would be best if the routers themselves cooperate and automate this subnetid configuration. This would allow rapid and ubiquitous IPv6 network deployment, and is even required in cases where the administration is impractical and users are not network-savvy. So in the rest of this paper, we will address this problem starting with a simple single-router topology in section 4 and extending it to complex multi-router environments in section 5.

4   Auto-configuring Single-Router IPv6 Networks

A single-router network is one, where a router connects several segments into a star topology as illustrated in Figure 2. A typical example of a single-router IPv6 networks is a simple home network, where the router, usually called a home or residential gateway, has several different internal segments such as Ethernet, WiFi (802.11), HomePNA, IEEE 1394, Bluetooth, etc., and also provides broadband Internet connectivity over one of xDSL, cable, or ISDN.

In Figure 2 the router connects 3 internal segments, S1, S2 and S3 together over its interfaces 1, 2 and 3; and provides Internet connectivity over its interface 4. Auto-configuring this network is quite easy: The router first gets the global routing prefix, g, from the ISP, then simply assigns unique 16-bit subnetids, a, b and c to internal segments S1, S2 and S3, i.e., prefixes g.a::/64, g.b::/64 and g.c::/64. The router then advertises these prefixes over the respective segments, and hosts would use IPv6 stateless auto-configuration algorithm of section 3 to

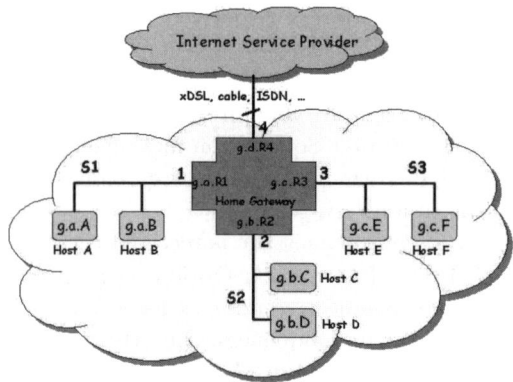

Fig. 2. An example single-router IPv6 network, e.g., a home network

auto-configure IPv6 addresses. In our example network, host A has configured IPv6 address g.a.A, where g is the 48-bit global routing prefix, a is the 16-bit router-assigned subnetid for the segment and A is the host's 64-bit interface-id.

In case the network is not connected to the Internet, e.g., an isolated network formed on-demand, the router would not get a global routing prefix, g, and must use local unicast addresses. We propose that the router first generates a 40-bit global ID using the algorithm in [7], and append it to FD00::/8 to make it a 48-bit local address prefix, $g1$. The router then assigns unique 16-bit subnetids, a, b and c to internal segments S1, S2 and S3, i.e., prefixes g1.a::/64, g1.b::/64 and g1.c::/64, and continues auto-configuration similar to the global prefix case. If a global routing prefix becomes available in the future, the router can continue using the same subnetid assignment by simply replacing the local address prefix $g1$ with g and advertise the newly generated prefixes, g.a::/64, g.b::/64 and g.c::/64. The router can either phase out the local unicast prefix $g1$ by advertising a prefix lifetime of zero, or can continue using it for local communication.

5   Auto-configuring Multi-router IPv6 Networks

A multi-router IPv6 network is one where two or more routers connect several segments together. One or more of the routers may also provide Internet connectivity. Any non-trivial network would have multiple routers.

Figure 3 depicts an example multi-router IPv6 network having 4 routers connecting 10 segments. In a typical deployment scenario, such as an engineering college, R1 would be deployed at a Computer Management Center and provide outside connectivity over xDSL or T1, and each internal router would be deployed at a different department, e.g., R2 at Computer Engineering, R3 at Electrical Engineering and R4 at Civil Engineering Departments. Departmental routers would also have connectivity between them for robustness and availability as in Figure 3. In a large organization, such as a big corporation or a university, the network may consist of tens of routers.

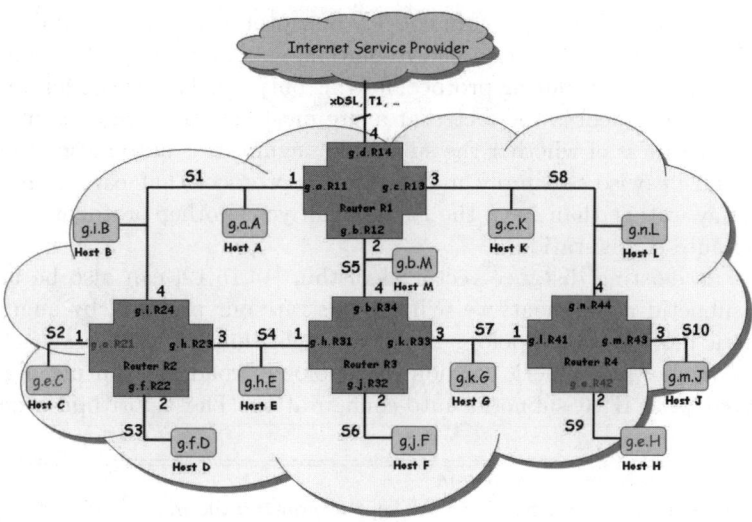

Fig. 3. An example multi-router IPv6 network, e.g., a college network

In this section, we propose extending the single-router auto-configuration algorithm of section 4 to multi-router networks. We assume that the network is connected to an ISP, and at least one border router obtains a 48-bit ISP-supplied global routing prefix denoted as g in Figure 3. We assume that this 48-bit global routing prefix is disseminated to all routers in the network using a mechanism such as an intra-domain routing algorithm. What is left for auto-configuration is the unique and consistent assignment of 16-bit subnetids to all segments of the network. To automate this process, our idea is to have the routers randomly assign locally unique subnetids to each of their interfaces during startup, and then cooperate with other routers to make their subnetid assignment unique over the entire network.

Since each router makes local subnetid assignment on their own, it is possible for 2 or more routers to assign the same subnetid over different segments, which we call a *subnetid conflict*. Subnetid conflicts can also occur when two operating networks are joined together by a hub or a bridge. Such subnetid conflicts must be detected and resolved for the network to function properly. For example, in Figure 3 we have one subnetid conflict: R1 has assigned IPv6 subnet g.e::/64 over S1, and R4 has assigned the same IPv6 subnet over S9. This conflict must be detected and one of the routers must change its IPv6 subnetid assignment to a network-wide unique value to resolve it. Note that assignment of the same IPv6 subnet over the same segment by different routers does not constitute a subnetid conflict, but this is in fact desirable, e.g., both R1 and R3 assign the same IPv6 subnet g.b::/64 to S5. Also notice that it is OK for two routers to assign different IPv6 subnets to the same segment, e.g., R1 assigns g.a::/64 to S1 while R2 assigns g.i::/64. What is problematic is the case when two or more routers assign the same IPv6 subnet over different segments.

While it is possible to design an entirely new protocol for subnetid dissemination, subnetid conflict detection and resolution, we simply propose extending an existing intra-domain routing protocol for this purpose. The reason for augmenting a routing protocol is the fact that a site must run an intra-domain routing protocol regardless of whether the subnetid assignment is automatic or manual. So it would be wise to simply make use of a protocol that each router must run anyway, rather than have the routers run yet another protocol for unique subnetid auto-configuration.

While an existing distance-vector algorithm [14,15,12] can also be used for unique subnetid assignment, we will demonstrate our proposal by augmenting the generic link-state or topology broadcast algorithm [16,17,13].

Figure 4 shows a generic version of topology broadcast routing algorithm augmented with IPv6 subnetid auto-configuration. The algorithm is designed

---

Symbols used in the algorithm:

s_n: Sequence number at node n (kept in non-volatile memory).

subnetid$_n$: subnetid of node n (kept in non-volatile memory).

N_n: Set of nodes neighboring to n.

$w_{n,m}$: Weight of link (n, m) for all $m \in N_n$.

$L_n = \{(m, w_{n,m}) : m \in N_n\}$.

$\mathcal{L}_n$: List of nodes known to n.

s_n^m: View of node n about s_m.

L_n^m: View of node n about L_m.

subnetid$_n^m$: View of node n about **subnetid**$_m$.

Augmented Topology Broadcast Routing Algorithm for a Node n:

I. Node n comes up:
    I.1. If (comes up first time) then
        I.1.1. $s_n \leftarrow 0$, **subnetid**$_n \leftarrow$ A unique subnetid.
    I.2. $N_n \leftarrow \emptyset$, $\mathcal{L}_n \leftarrow \{n\}$.
    I.3. Bring up all adjacent operating links.

II. Adjacent link (n, m) goes down:
    II.1. Delete m from N_n.

III. Adjacent link (n, m) comes up:
    III.1. $w_{n,m} \leftarrow$ measured weight for (n, m).
    III.2. Add m to N_n.

IV. Periodically:
    IV.1. $s_n \leftarrow s_n + 1$.
    IV.2. Send $(n, s_n, \textbf{subnetid}_n, L_n)$ to all neighbors in N_n.

V. Node n receives message $(m, s, \textbf{subnetid}, L)$:
    V.1. If $(m \notin \mathcal{L}_n$ or $s_n^m < s)$ then
        V.1.1. If $(m \notin \mathcal{L}_n)$ then Add m to $\mathcal{L}_n$.
        V.1.2. $(m, s_n^m, \textbf{subnetid}_n^m, L_n^m) \leftarrow (m, s, \textbf{subnetid}, L)$.
        V.1.3. If $(\textbf{subnetid} = \textbf{subnetid}_n$ and $m \notin N_n$ and $m < n)$ then
            V.1.3.1. **subnetid**$_n \leftarrow$ A new subnetid. /* Conflict */
        V.1.4. Send $(m, s, \textbf{subnetid}, L)$ to all neighbors in N_n.

Fig. 4. Topology Broadcast Routing Algorithm for IPv6 Router Auto-configuration

for a single interface (designated as node n) of a router. So a router will run an instance of the algorithm for each of its interfaces. In steps I.1 and I.2, a router simply assigns a locally-unique subnetid to the interface during startup. Then steps IV.1 and IV.2 are the periodic link-state broadcast, which diseminates the neighbor list and the router-assigned subnetid for the link (segment) to other routers in the network. Steps V.1 through V.4 are where the router performs subnetid conflict detection and resolution: When the router receives a link-state packet from non-neighbor router interface m, i.e., a router interface not attached to the same link as n, with subnetid $subnetid_m$, it checks for a subnetid conflict in step V.1.3. In the case of a subnetid conflict, i.e., $subnetid_n = subnetid_m$, the router with the bigger identifier changes its subnetid to resolve the conflict at step V.1.3.1, i.e., the router with the smaller identifier wins the conflict battle. Since the next message sent by the loser node will have a bigger sequence number due to step IV.1, all nodes in the network will start using the new subnetid.

It can be shown that given a multi-router network having k router interfaces, if no further topological changes occur, then the network enters a state where no IPv6 subnetid conflicts exist and each router has established a loop-free path to all IPv6 destinations, after $O(k \times e)$ message exchanges, where e is the total number of segments in the network. Informal proof is as follows: Observe that the router interface having the smallest identifier will win every IPv6 subnetid conflict battle, if there is any. So after a message from this node reaches all nodes in the network, which takes e messages, all nodes in the network would have detected and resolved any IPv6 subnetid conflicts with this node and the IPv6 subnetid assigned to this smallest identifier node will not change after that. Next the node with the second smallest identifier would stabilize its IPv6 subnetid within the network in e messages and so on. Lastly the node with the maximum identifier will have its IPv6 subnetid fixed. So irrespective of the number of IPv6 subnetid conflicts in the initial state of the network, they would all be resolved within $O(k \times e)$ message exchanges. Additionally, by the end of $O(k \times e)$ messages, all routers would have learned the complete topology of the network and can easily compute shortest routes to all IPv6 destinations.

Probability of subnetid conflict: For a network with n router interfaces, and L-bit subnetid value, the probability of collision can be computed using the formula $p = 1 - e^{\frac{-n^2}{2^{L+1}}}$ given in [7]. Given L=16, and assuming n = 20, $p = 3.05 \times 10^{-3}$.

Lack of a global routing prefix: In case the network is not connected to the Internet, the site would lack a global routing prefix, and local addresses must be used. Two choices exist: (1) Have each router generate its own 48-bit prefix and advertise it over the network. With n routers in the network, we would have n different prefixes within the site, (2) Have routers agree upon a single 48-bit local prefix and use it for the entire site.

In the first case, each router i simply generates its own 48-bit global ID prefix, g_i before making locally unique 16-bit subnetid assignment to each of its interfaces. Thus router i with k interfaces would assign unique 64-bit prefixes

$g_i.s_j::/64$, $1 \leq j \leq k$ to each of its j interfaces. These 64-bit prefixes would almost always be unique within the site. In [7] authors show that if 10 different 40-bit prefixes are generated within a site using the 40-bit global ID field in a local unicast address, the probability of collision is 4.54×10^{-11}. This implies that with this auto-configuration method no IPv6 subnet conflicts will occur within the site, so the problem will simply boil down to distributing these unique prefixes to all routers using a routing protocol such as the one in Figure 4. If a global routing prefix, g, becomes available in the future, each router can simply generate new 64-bit prefixes $g.s_j::/64$ for each of its j interfaces, and run the algorithm in Figure 4 to advertise its global IPv6 subnets, and also detect and resolve potential IPv6 subnetid conflicts.

In the second case, each router i generates its own 48-bit global ID prefix, g_i, and assigns locally unique 16-bit subnetids to each of its interfaces. But this time, routers not only agree upon consistent 16-bit subnetid assignment, but also upon a single 48-bit local prefix to be used for the entire site. This single prefix selection can simply be incorporated to the algorithm in Figure 4. At step V, when a router receives a link-state advertisement, it first enters into a 48-bit prefix election. Similar to subnetid conflict resolution, we can simply let the router having the smaller identifier win the prefix battle. Thus when the network reaches the stable state, the 48-bit prefix generated by the router having the smallest identifier will be elected as the single site identifier. 16-bit subnetid conflict detection and resolution will then be the same as in the algorithm in Figure 4. The advantage of this approach is the following: Since the network will use a single 48-bit local prefix, 16-bit subnetids would uniquely be assigned over all segments. So when a 48-bit global prefix becomes available, routers can simply use their current 16-bit subnetid assignments to generate 64-bit prefixes using the global prefix. Thus the same 16-bit subnetids would be used by both global prefixed addresses and local prefixed addresses as recommended in [7].

6   Hybrid Networks

Although it is possible to have the entire IPv6 network auto-configured as described in section 5, it is usually desirable to manually configure some of the subnets. This is necessary in such cases where some of the servers in the network need to have well-known, fixed IPv6 addresses.

Figure 5 shows the high-level organization of Anadolu University's internal network. The university has a Web Server, a DNS Server, a Mail Server and an FTP server all attached to segment S1 of the edge router R1, that must have fixed, well-known IPv6 addresses. This means that the administrators want segment S1 to have a fixed subnetid. In the example the administrators have configured IPv6 subnet g.a::/64 over S1. Similarly some departments within the university, which may have their own skilled administrators, such as the computer engineering department represented by router R2, may want to have their own Web Server, FTP Server etc. So they may want to administer some of the segments of their own router (such as segment S3 in the example network). Other

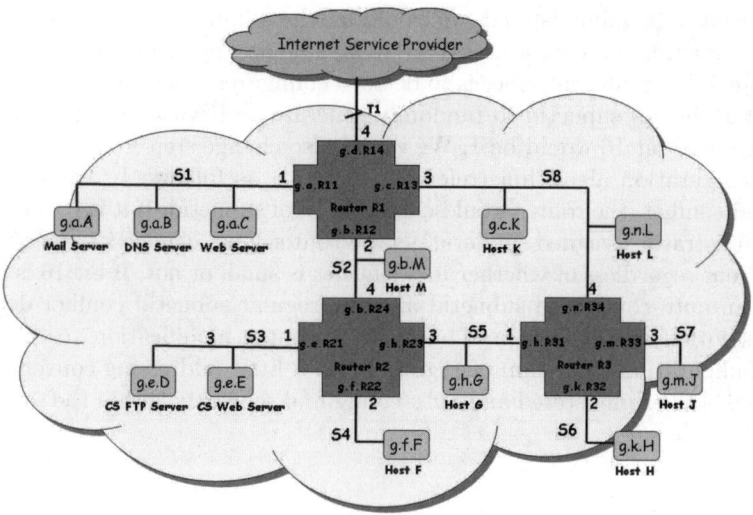

Fig. 5. An example multi-router IPv6 network with segments S1 and S3 administered, and the rest auto-configured

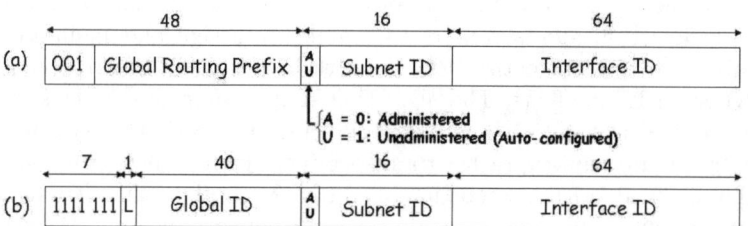

Fig. 6. (a) Global, (b) local unicast address structures with administered(A)/ unadministerd(U) flag within subnetid to define administered or auto-configured IPv6 subnets within a site

departments may not have the necessary technical skills or may not be interested in configuring their internal networks, and simply leave it to the routers to auto-configure themselves. Therefore, the rest of the segments are left unadministered and are auto-configured by the routers.

In such hybrid networks, the question then arises as to how we can have both administered and auto-configured segments within the same IPv6 network? We propose the following solution to this problem: Let the most significant bit (bit 16) of the subnetid represent whether the subnetid is administered or auto-configured. If this bit is 0 then the subnetid is assigned by an administrator, if this bit is 1 then the subnetid is auto-configured by a router (refer to Figure 6). Notice that the first 48-bits of the prefix still either represent the global routing prefix or the local prefix depending on whether the network is connected to the Internet or not (refer to Figure 6). With this convention, if a router interface would need to be administered, we would let the administrators use the last 15-bits of subnetid to create a unique IPv6 subnet for the router interface, bit 16 would be 0. If a router interface is to be auto-configured, then the router still uses the last 15-bits of subnetid to randomly generate an IPv6 subnet for the router interface, and bit 16 would be 1. We would also change step V.1.3 of our router auto-configuration algorithm (refer to Figure 4) as follows: In the event of a subnetid conflict, the router would check bit 16 of subnetid: If it is 0, then this is an administrator-assigned subnetid, so the router must change its own subnetid assignment regardless of whether its identifier is small or not. If bit 16 is 1, then this is an auto-configured subnetid and the regular subnetid conflict detection and resolution would kick in. Thus with a simple modification to the router auto-configuration algorithm of Figure 4 and a little addressing convention, we can have both administered and auto-configured segments within the same IPv6 network.

7   Concluding Remarks

In this paper we proposed an IPv6 router auto-configuration algorithm that in combination with IPv6 host auto-configuration protocol allows complex multi-router IPv6 networks to be fully auto-configured without manual intervention. Proposed algorithm allows routers to seamlessly assign and maintain unique IPv6 subnets within the entire IPv6 network by detecting and resolving potential IPv6 subnetid conflicts. The algorithm simply augments an existing intra-domain routing protocol such as RIP [12] or OSPF [13] that every multi-router network must run anyway, rather than specifying yet another protocol. We also showed how the algorithm can be modified to allow both administered and auto-configured segments to co-exist within the same network together. We claim that the addition of such an auto-configuration algorithm to routers would greatly simplify deployment and management of IPv6 networks both in big organizations where some of the segments may need to be administered, and in small networks such as home and SOHO networks, where the entire network may need to be auto-configured.

References

1. Microsoft Corporation, "Introduction to IP Version 6", http://www.microsoft.com/windowsserver2003/technologies/ipv6/introipv6.mspx, September 2003.
2. S. Thomson and T. Narten, "IPv6 Stateless Address Autoconfiguration", RFC 2462, December 1998.
3. T. Narten, "Neighbor Discovery and Stateless Autoconfiguration in IPv6", IEEE Internet Computing, pp. 54-62, August 1999.
4. D. Binet, "Home Networking: The IPv6 killer application?", http://aristote1.aristote.asso.fr/Presentations/IPv6-2002/P/Binet/Binet.pdf, September 2002.
5. R. Hinden and S. Deering, "IP Version 6 Addressing Architecture", RFC 4291, February 2006.

6. R. Hinden, S. Deering and E. Nordmark, "IPv6 Global Unicast Address Format", RFC 3587, August 2003.
7. R. Hinden and B. Haberman, "Unique Local IPv6 Unicast Addresses", RFC 4193, October 2005.
8. O. Troan and R. Droms, "IPv6 Prefix Options for Dynamic Host Configuration Protocol (DHCP) version 6", RFC 3633, December 2003.
9. S. Miyakawa and R. Droms, "Requirements for IPv6 Prefix Delegation", RFC 3769, June 2004.
10. A. Thulasi and S. Raman, "IPv6 Prefix Delegation Using ICMPv6", draft-arunt-prefix-delegation-using-icmpv6-00.txt, April 2004.
11. C. Huitema and B. Carpenter, "Deprecating Site Local Addresses", RFC 3879, September 2004.
12. G. Malkin, "RIP Version 2", RFC 2453, November 1998.
13. J. Moy, "OSPF Version 2", RFC 2328, April 1998.
14. J. M. Jaffe and F. H. Moss, "A Responsive Distributed Routing Algorithm for Computer Networks", IEEE Transactions on Communications, vol. COM-30, number 7, pp. 1758-1762, July 1982.
15. P. M. Merlin and A. Segall, "A Failsafe Distributed Routing Protocol", IEEE Transactions on Communications, vol. COM-27, number 9, September 1979.
16. P. A. Humblet and S. R. Soloway, "Topology Broadcast Algorithms", Computer Networks and ISDN Systems, vol. 16, pp. 179-186, 1989.
17. J. M. Spinelli and R. G. Gallager, "Event Driven Topology Broadcast Without Sequence Numbers", IEEE Transactions on Communications, vol. 37, number 5, May 1989.

An Efficient Process for Estimation of Network Demand for QoS-Aware IP Network Planning

Alan Davy, Dmitri Botvich, and Brendan Jennings

Telecommunications Software & Systems Group,
Waterford Institute of Technology,
Waterford, Ireland
{adavy, dbotvich, bjennings}@tssg.org

Abstract. Estimations of network demand are an essential input to the IP network planning process. We present a technique for per traffic class IP network demand estimation based on harnessing information gathered for accounting and charging purposes. This technique represents an efficient use of pre-existing information, is easy to deploy, and, crucially, is highly cost-effective in comparison to traditional direct measurement systems employing dedicated traffic metering hardware. In order to facilitate QoS-aware network planning we also introduce a technique for estimation of QoS related effective bandwidth coefficients via analysis of a relatively small number of packet traces. The combination of the demand and effective bandwidth coefficient estimation techniques provide the basis for an effective, low-cost network planning solution. In this paper we present initial results that validate our contention that network accounting records can be reused to create a QoS aware demand matrix for IP networks.

1 Introduction

An ISP must ensure that network resources are being utilised optimally in order to avoid unnecessary expenditure. Of course, it is also important that the process of planning, deploying and managing network resources is itself cost-effective and does not significantly degrade network performance. Network planning, in particular, is a vital part of any ISP's business. Currently, network planning typically involves the use of dedicated metering hardware to gather and collate large amounts of network activity data, which is then used to identify an optimal network configuration design reflecting estimated demand. Use of dedicated hardware means that this approach is relatively expensive, incurring costs in hardware procurement, depreciation and maintenance, as well as significant training and operational costs. In this paper we outline the foundations of an efficient, QoS-aware network planning process, that, through re-use of networking accounting data, can be delivered at relatively low cost and with minimal impact on network performance. We contend that ISPs can easily reuse traffic data gathered by network accounting systems to generate sufficiently accurate estimations of network demand. In this paper we show that such reuse of accounting data to construct a QoS-aware demand matrix results acceptable relative error for large amounts of traffic. We also present a light-weight technique for

G. Parr, D. Malone, and M. Ó Foghlú (Eds.): IPOM 2006, LNCS 4268, pp. 120–131, 2006.
© Springer-Verlag Berlin Heidelberg 2006

estimation of QoS aware effective bandwidth coefficients, which in conjunction with the QoS-aware demand matrix provides the basis for an effective, low cost QoS-aware network planning process.

The paper is organised as follows. §2 disuses related work in the area of network accounting and network planning. §3 provides a description of our architecture, detailing the components and tools used to gather and analyse traffic data. We present our algorithm for calculating the demand matrix from accounting records in §4. §5 discusses how we plan to take QoS into consideration by estimating effective bandwidth coefficients. We validate our architecture in §6 through the use of a specified scenario and a prototype test bed. Finally we evaluate our results in §7 and conclude with future work in §8.

2 Related Work

The concept of network planning for QoS aware network optimisation has been put forward by Wu and Reeves [1]. They look at capacity planning in a DiffServ network, and focus on a network with two traffic classes, Expedited Forwarding (EF) and Best Effort. Their work focuses on developing an optimisation algorithm that jointly, selects a route in the network for each EF user demand (Origin - Destination) pair, and assigns a capacity value for each link within the network to minimise the total link cost, subject to the performance constraints of both EF and BE traffic classes. Our work intends on developing a network planning solution to take into consideration all DiffServ traffic classes.

The demand matrix has been associated with a wide range of network planning activities such as network design, traffic engineering and capacity planning [1]. The demand matrix has been shown to be an effective method of representing network wide demand on the network from edge to edge [2, 3, 1, 8]. There are a number of different approaches in calculating the demand matrix, the major concern being whether to calculate the demand matrix from direct measurement [3], or to use summarised sampling methods such as trajectory sampling [9]. The algorithm we propose is a centralised approach to calculating the demand matrix. This is primarily based on the work of Feldman et al [3]. Our algorithm varies as it is limited to the types of metering records used by accounting systems. For efficiency processing may also be distributed further out onto the edge nodes themselves [4].

The IETF have developed a number of network accounting architectures such as RADIUS [11], and DIAMETER [12]. These systems rely on the collection of metering information from the network, and forward this information to mediation points following a particular format. These systems are most commonly used for VoIP accounting and other such session based services. The IP Multimedia Subsystem has defined DIAMETER as its accounting protocol of choice [13]. Cisco have developed NetFlow [14] as their metering and accounting system. NetFlow is widely used in the industry for various operations such as IP network accounting and billing, user and application monitoring, network planning, security analysis and traffic engineering. The IETF have recognised this industry standard and have developed the IETF IP Flow Information Export (IPFIX) [1] architecture based on it. We have based our accounting system on the IPFIX architecture.

3   QoS-Aware Network Planning Architecture

The architecture illustrated in Fig. 1. extends the traditional network accounting architecture to facilitate construction of QoS-aware demand matrices and for the calculation of QoS-aware effective bandwidth coefficient from collected packet traces.

All network accounting systems depend on metering information to account for service usage within their network domain. The network accounting systems capture summarised information from the network in the form of flow records, of the form depicted in below.

Src Address	Src Port	Dest Address	Dest Port	Protocol	TOS	Packets	Size	Start time	Active	Idle

A flow record represents a set of IP packets passing a network interface that possess common properties, such as the same source address, source port, destination address destination port and protocol. A flow record can usually be associated with unidirectional traffic of a particular application session, such as a VoIP call. An example of network accounting systems that use flow records are Cisco Netflow [14], and the IETF proposed standard IPFIX [1]. These records are used as a base for rating usage of traffic within the network. Once this traffic is rated, it can be associated with user sessions where the users can be charged and billed for service usage. We intend to use the flow records collected by accounting systems to construct a view of the traffic demand on the network, other wise known as a demand matrix.

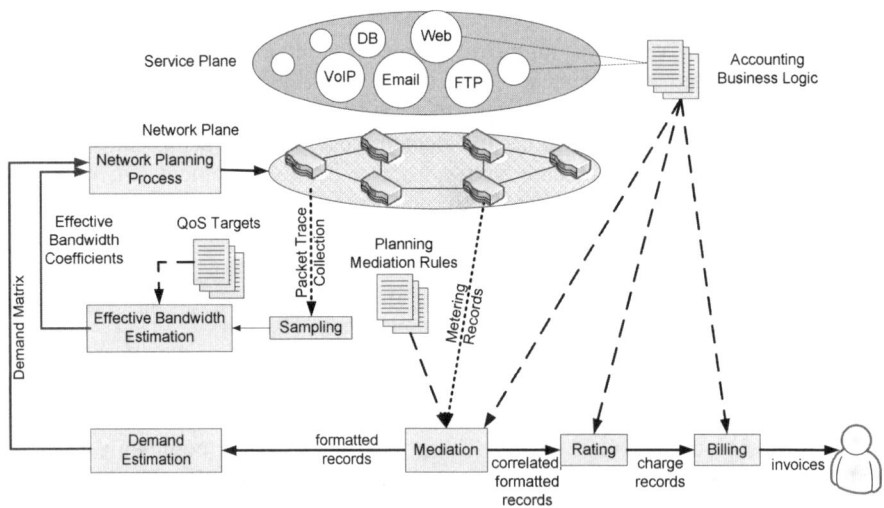

Fig. 1. Network Accounting and Planning Architecture

The demand matrix captures network wide demand from information collected at ingress and egress edge routers. We assume that all traffic entering at an ingress edge node, must exit at an egress edge node, i.e. no traffic is consumed within the core. If this is the case, we can build a picture of total traffic demand within the network between edge nodes. The demand matrix stores this information as a pair-wise edge-to-edge matrix. To create this matrix, we must have knowledge of where traffic exits the core network, and match this against its ingress. We wish to build a demand matrix for each particular traffic class on the network as to associate particular traffic class demand characteristics with QoS targets within the accounting system. This is achieved by adding an additional dimension to the demand matrix matching the *Type of Service* (TOS) field of metered traffic.

The TOS field identifies the QoS traffic class of a particular IP packet. This information is carried on into the flow records. The DiffServ QoS architecture uses the TOS field to assign DiffServ Code Points (DSCP) to IP packets to include them into particular traffic classes. QoS levels for these traffic classes can then be controlled within the network, giving the traffic class particular loss, delay, jitter, and throughput targets. By adding the TOS dimension our demand matrix becomes QoS-aware.

The *Mediation* component exports collected metering records in the required format to the *Demand Estimation* component. This process will construct a multi-dimensional demand matrix of all accounted traffic passing through the network from edge to edge and per traffic class. Our observation is that accounting systems record a considerable amount of network activity, while at the same time summarising this information to reduce processing and storage. We contend that using existing metered information to estimate the demand matrix is less expensive than calculating the demand matrix through the use of dedicated hardware using direct measurement methods. Of course, this data may not provide an accurate measurement of network demand, however, as shown below, the relative error introduced is unlikely to impact significantly on the efficacy of the planning process.

A relatively small number of packet traces are taken from the network and analysed to calculate appropriate bandwidth thresholds known as effective bandwidth to be reserved in order for this traffic to maintain defined QoS targets. The process of packet trace collection is a very light weight approach to sampling network activity over a short interval of time, which has little effect on network resources.

Once the demand matrix and effective bandwidth coefficients are estimated, they are used as input to the *Network Planning* component, which will use the collected and analysed information to develop an optimised network configuration that will ensure imposed network QoS targets are maintained.

4 Demand Matrix Estimation Process

Accounting systems mediate metering records to associate service usage with users. Similar to this approach we need to associate the metering records with edge to edge network demand. We outline our algorithm to achieve this in Fig. 3. The algorithm calculates the demand on the network a particular flow has, per interval. This is necessary as the demand matrix looks at total network demand per interval of time, e.g. 10 min intervals.

Fig. 2 depicts an algorithm of estimating network demand from flow records per interval. Each flow record will have a start time (t^f_{start}) and an end time (t^f_{end}). The flows rate r^f can be calculated from the flow size, stored in the flow record, divided by the flow duration. The diagram shows 4 cases the algorithm captures. The objective of the algorithm is to sum up all demand of all flows that lie within a particular time period $\{t_n, t_{n+1}\}$. Case 1 captures demand of flows that end within the time period. Case 2 captures demand of flows that start within the time period. Case 3 captures demand of flows that start and end within the time period, and finally Case 4 captures demand of flows that are active through the whole time period. This algorithm makes the assumption that packet arrival time within the flow is of a uniform distribution. By taking this assumption the proportion of demand can be calculated by multiplying the flow rate by the duration of time the flow exists within the current interval. This will lead to some inaccuracy in the final value, as throughout the flow's duration, packet distribution is not normally uniform. To calculate the complete network wide demand matrix all metering records collected from on all metering devices on all ingress edge routers are processed.

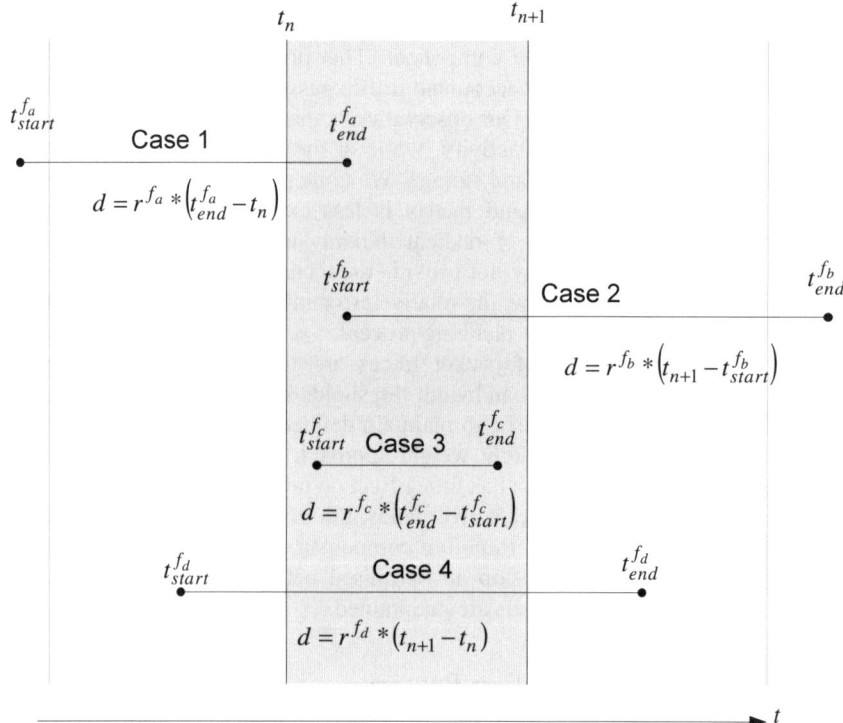

Fig. 2. Estimation of flow demand per interval of time

The algorithm expressed in Fig.3 has five nested for loops, looping through each ingress router, each metering device, and each source node hanging off that metering device, each interval in time, and each flow record within that current metering device. The algorithm matches each flow record to a source node, and estimates the flow's demand within the current interval. The algorithm then matches the destination address of the flow record to a particular egress edge node. This mapping allows us to identify where the traffic is exiting the network.

```
Input: (EdgeRouters, interval, TotalTime  )
Output: (DemandMatrix[IngressRouter, EgressRouter, TOS] )
For each ingressRouter in EdgeRouters
    For each meteringDevice in edgeRouter
        For each sourceNode of meteringDevice
            For each timeInterval in TotalTime
                For each flow of meteringDevice
                    If flow_src_ipaddr == sourceNode
                        Demand =
                                calculateDemand(timeInterval,
                                interval, flow)
                        egressRouter =
                                findEgressRouter(flow_dest_ipaddr )
                        DemandMatrix[ingressRouter, egressRouter,
                                flow_TOS ] += Demand
    Return DemandMatrix
```

Fig. 3. Demand Matrix Estimation Algorithm

The *findEgressRouter* function in Fig. 3b is used to return the egress router node corresponding to where the flow exits the core network. This function is a simple static table lookup at the moment but can be extended to retrieve this value dynamically through BGP table lookups. Once the egress edge node is found, an entry is added to the demand matrix. Once all records have been processed the demand matrix is returned.

5   Taking QoS Requirements into Account

The network planning process requires a view of demand on the network. We propose estimating network demand by the generation of a demand matrix from accounting records. We wish to develop a network planning scheme that will provide improved QoS guarantees. To take QoS targets into account per traffic class, we propose a light weight method of estimating the effective bandwidth of a traffic class, between edge nodes.

There are a number of definitions of what effective bandwidth is, for example see [4, 5]. From the different methods of estimating the effective bandwidth, practically all of them are based on building a traffic model. The building of an accurate traffic model for a bursty traffic source is quite a challenging task. In particular, it is quite difficult to take into account the different activity levels of a traffic source for different time scales ranging from milliseconds to minutes and hours.

The effective bandwidth of a traffic source is a minimal link rate which can guarantee certain specified QoS targets. Effective bandwidth can be defined for different types of QoS targets including delay, loss or both delay and loss targets together. In this paper we are interested in QoS delay targets only. A QoS delay target specifies the maximum delay experienced on the network and the proportion of traffic which is allowed exceed this maximum delay.

A typical example of a QoS delay target is (50ms, 0.001) which means that only 0.1% of traffic is allowed to be delayed more than 50 ms. As effective bandwidth depends on the QoS target; for different QoS targets, effective bandwidth could be different.

Our approach of estimating effective bandwidth follows a more empirical method and fits well to our main objective related to QoS aware network planning. A brief description of the algorithm we wish to implement is as follows. Suppose the QoS delay target ($delay_{max}$, p_{delay}) is fixed and includes $delay_{max}$ the maximum delay and p_{delay} the percentage of traffic which can exhibit delay more than $delay_{max}$. We define effective bandwidth R_{eff} of a traffic source for delay QoS target ($delay_{max}$, p_{delay}) as a minimal link rate such that if we simulate a FIFO queue (with unlimited buffer) the percentage of traffic which will exhibit delay more than $delay_{max}$ will be less than p_{delay}. We will assume that initially the queue is empty.

To estimate the effective bandwidth of a particular traffic source on the network, we take a recorded packet trace of that source. The algorithm we define for estimating the effective bandwidth of a recorded packet trace is as follows. The algorithm is based on the following observations. Suppose we simulate a FIFO queue with the same traffic source for different queue rates $R_1 > R_2$ and estimate the percentages p_1 and p_2 of traffic delayed more than $delay_{max}$ for different rates respectively, then $p_1 \leq p_2$. This means that the percentage of traffic delayed more than $delay_{max}$ is a monotonically decreasing function of the queue rate. Using this observation it is straight forward to design an algorithm for a recorded packet trace to find the minimal value of a queue rate such that the percentage of traffic delayed more than $delay_{max}$ is

$$k_i = \frac{R_{eff,i}}{mean_i}$$

less than p_{delay}.

We assume that the QoS delay target ($delay_{max}$, p_{delay}) is fixed per traffic class. We take a large number of recorded traffic traces of more or less the same duration T_{max}. The choice of T_{max} is important. If T_{max} is too large or too small, the estimated ratio of effective bandwidth to the mean rate will be underestimated. Typically T_{max} is chosen between 1 minute and 1 hour, e.g. 10 minutes. Suppose we have N traffic traces. For the ith traffic trace we estimate both $R_{eff,i}$ and $mean_i$ and calculate

We note that the effective bandwidth is always larger or equal to the mean rate. So for all i, $k_i \geq 1$. We now consider a set of N effective bandwidth coefficients $\{k_1,..., k_N\}$. First we exclude any k_i with too small a mean rate using some appropriate threshold value. Second we calculate K_{95} the 95th percentile. The effective bandwidth coefficients K_{95} is used for our purposes.

6 Experimental Evaluation

Fig 4. illustrates an ISP providing connectivity and services to a number of customer groups. A customer group defines a set of service interactions with a number of services offered by the ISP to fulfil a particular business process. The ISP offers various levels of QoS guarantees to the customers depending on the type of service the customer requests and the amount of revenue generated by that service interaction. The ISP therefore has a set of QoS targets it must maintain within its network in order to generate maximum revenue.

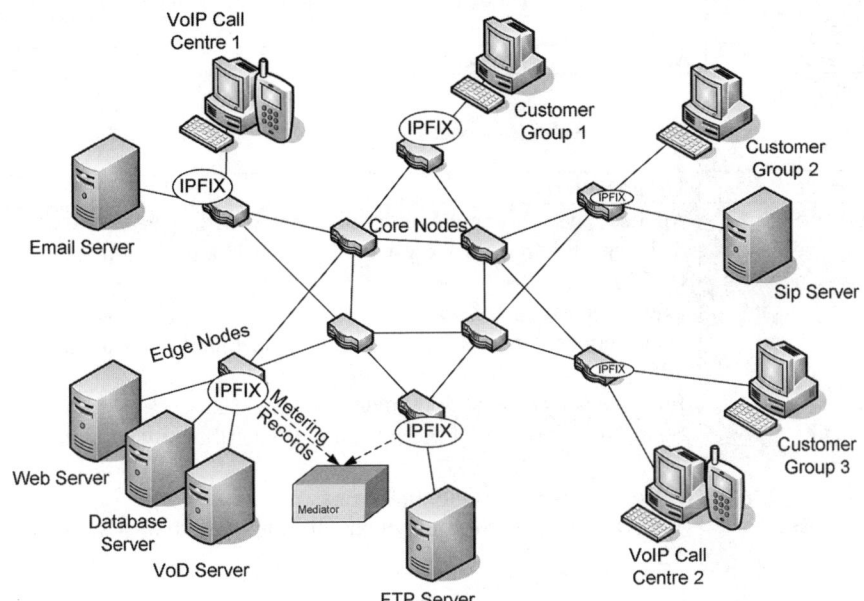

Fig. 4. Scenario Topology

The ISP has a number of application servers distributed through the network offering services to the customers. The customer is located at a number of customer group LANs distributed throughout the network. The ISP has an accounting system in place to meter service usage. Metering information is collected at key points throughout the network through the use of strategically positioned metering devices.

This metering information is then used for rating and billing purposes by the service providers accounting system. The initial step of our proposed network planning process utilises this accounting information to estimate the demand imposed on the network by service interactions.

To test and validate our proposed algorithms in Fig. 3 we have implemented a use case based on the scenario with a prototype network topology. We use OPNET as our network simulation environment. We compare our approach of calculating the demand matrix to direct measurements taken, and show that our approach produces acceptable relative error. The network topology has four core routers, and five edge routers, each connected by 10Mbps Ethernet. All IP traffic is generated at the edge of the network, no IP traffic is consumed at the core of the network.

There are three customer groups each containing 25, 35 and 50 customers respectively. There are four services the service provider offers to the customers on the network. They are email, FTP, database and VoIP (voice over IP). These are

Table 1. Traffic Pattern Setup

Service	DSCP	Pattern 1	Pattern 2	Pattern 3
SIP VoIP	EF	5 users	15 users	30 users
FTP	AF21	5 users	5 users	5 users
Email	AF31	10 users	10 users	10 users
Database	AF41	5 users	5 users	5 users

Table 2. Service Characteristics

Service	Usage pattern
SIP VoIP	Silence length is exponentially distributed with a mean of 0.65s Talk spurt length is exponentially distributed with a mean of 0.352s Encoding rate is 8 Kbps
FTP	Inter request time is exponentially distributed with a mean of 720s File size is a constant 5000 bytes
Email	Send / receive interval is exponentially distributed with a mean of 720s Email size is a constant 3000 bytes
Database	Transaction interval is exponentially distributed with a mean of 12s

located on application servers distributed around the edge of the network. An IPFIX device has been modelled in OPNET and attached to each ingress interface of the edge nodes. All traffic entering the network is available for both direct measurement and accounting based metering. Each of the four services is set up to generate a particular traffic pattern within the network for customer group 2, see Table. 1. Customer group 1 and 3 will have set traffic patterns for all three simulations. For each simulation Customer group 2 will follow a different traffic pattern; each traffic pattern will have an increased number of VoIP users on the network, thus increasing the amount of EF traffic generated across two particular edge nodes. Each customer within a group is set up to interact with the offered services following a particular service usage pattern, outlined in Table 2.

7 Experimental Results

We ran the OPNET simulation over a 2 hour simulation period, collecting direct measurements and accounting records from all ingress interfaces of the edge nodes. Fig. 5 shows demand across edge 2 to edge 6 for traffic class EF. Fig.5a. shows the direct measurement of demand calculated from direct measurements. This graph has the highest resolution of accuracy. Fig.5b-d show the same edge to edge demand for the same traffic class, but are generated from the accounting records. In each case the interval over which the demand is calculated increases from 3 to 30, to 100 seconds. This means the demand values calculate for the demand matrix are over these three interval steps. The figures above show slight loss in accuracy by reducing the size of the sampling interval.

We calculate average demand over 10 min intervals for the three traffic patterns outlined in Fig. 6. By this we mean from the values held in the demand matrix, we calculate demand over 10 minute intervals, and compare these values to direct measurements. Traffic pattern 1 has the lowest demand over the simulation duration between edge 2 and edge 6 for traffic class EF. Traffic pattern 2 has a slightly larger demand across the edges and traffic pattern 3 has the largest demand across the two edges.

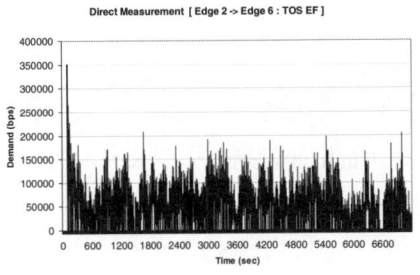

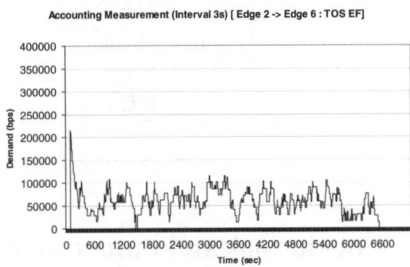

(a). Direct Measurement   (b). Accounting : Sample Interval 3s

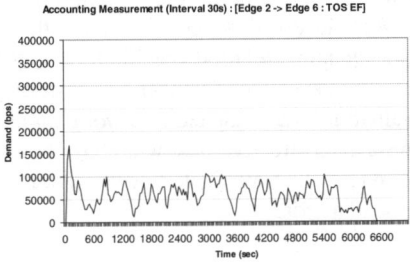

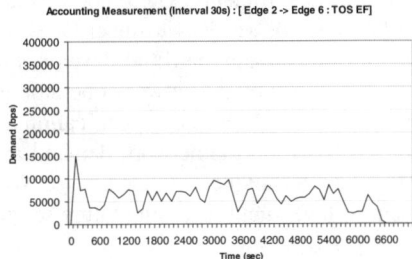

(c). Accounting : Sample Interval 30s   (d). Accounting : Sample Interval 100s

Fig. 5. Direct measured demand vs. accounting based demand

We take three traffic patterns and run them for a period of 2 hours. We calculate relative error in calculating demand based on accounting records in comparison to their associated direct measurements over 10 minute intervals. From these experiments we observe that our approach estimates demand with relative error of approximately 10 %. This can be tied to the fact the our demand estimation algorithm is based on analysing accounting records, which are in turn a summary of network traffic. As network planning is predominantly based on estimation of current network demands, a high level of accuracy is generally not required as future traffic demands are dependent on human usage trends, which are in themselves unpredictable. Therefore margin of error in demand estimation is quite acceptable for the purpose of QoS aware network planning, of which the demand matrix is a vital part of.

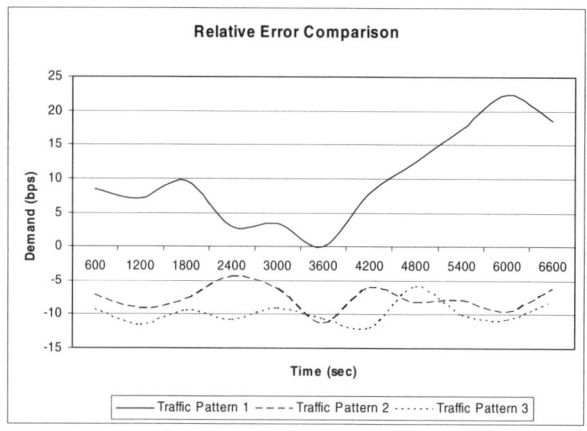

Fig. 6. Relative error across all traffic Patterns

8   Conclusions and Future Work

We proposed a method of estimating network demand from pre-existing flow records used by network accounting systems, and a method of calculating QoS related effective bandwidth coefficients. These coefficients tell us how much bandwidth is required per traffic class for services to meet QoS targets. We can use these coefficients with the QoS aware demand matrix to develop a network planning solution specific to the QoS targets outlined by the network operator and between the service provider and customer. Future work will focus on specification and evaluation of a complete network planning process based on the network demand and QoS-aware effective bandwidth coefficient estimation techniques outlined here. We also intend to investigate incorporation of business level input regarding future service demand trends.

Acknowledgements

The authors would like to take this opportunity to thank the anonymous referees for their useful comments and feedback on the paper. This work has received partial support from the following: M-Zones is funded under the HEA PRTLI, Science Foundation Ireland under the Autonomic Management of Communications Networks and Services programme (PI Cluster Award 04/IN3/I404C).

References

1. K. Wu and D. S. Reeves, "Capacity Planning of DiffServ Networks with Best-Effort and Expedited Forwarding Traffic," Telecommunications Systems, vol. 25, no. 3/4, pp. 193-207, (2004).
2. G. Sadasivan, N. Brownlee, B. Claise, and J. Quittek, "Architecture for IP Flow Information Export (Internet-Draft)," draft-ietf-ipfix-architecture-11.txt, http://www.ietf.org/html. charters/ ipfix-charter.html (2005).

3. A. Feldman, A. Greenberg, C. Lund, N. Reingold, J. Rexford, and F. True, "Deriving traffic demands for operational IP networks: methodology and experience," IEEE/ACM Transactions on Networking (ToN), vol. 9, no. 3, pp. 265-280, (2001).

4. K. Papagiannaki, N. Tatf, and A. Lakhina, "A Distributed Approach to Measure Traffic Matrices," Proceedings of the 4th ACM SIGCOMM Conference on Internet Measurement (2004).

5. F. Kelly, "Notes on effective bandwidth," in Royal Statistical Society Lecture Notes Series, 4. Oxford University Press, pp. 141-168,(1996).

6. J. Y. Hui, "Resource allocation for broadband networks," IEEE Journal on Selected Areas in Communications, vol. 6, pp. 1598-1608, (1988).

7. D. D. Botvich and N. Duffield, "Large deviations, the shape of the loss curve, and economies of scale in large multiplexers," Queueing Systems, vol. 20, pp. 293-320, (1995).

8. A. Medina, C. Fraleigh, N. Tatf, S. Bhattacharya, and C. Doit, "A Taxonomy of IP traffic matrices," Proc. International Society for Optical Engineering (SPIE) Vol. 4868, p. 200-213, (2002).

9. A. Soule, A. Lakhina, N. Tatf, K. Papagiannaki, K. Salamatian, A. Nucci, M. Crovella, and C. Doit, "Traffic matrices: balancing measurements, inference and modeling,", Joint International Conference on Measurement and Modeling of Computer Systems ACM Press New York, NY, USA, pp. 362-373, (2005).

10. N. Duffield and M. Grossglauser, "Trajectory sampling for direct traffic observation," New York, NY, USA: ACM Press, pp. 271-282,(2001).

11. C. Rigney, "RADIUS Accounting," in IETF RFC 2139 Internet Engineering Task Force, www.ietf.org/rfc/rfc2139.txt, (1997).

12. Calhoun P., Loughney J., Guttman E., Zorn G., and Arkko J., "Diameter Base Protocol," in RFC 3588 Internet Engineering Task Force, www.ietf.org/rfc/rfc3588.txt, (2003).

13. 3GPP2, "IP Multimedia Subsystem - Accounting Information Fows and Protocols," in All IP core network multimedia domain, 3GPP2 X.S0013-008-0, (2005).

14. Cisco Systems, "Introduction to Cisco IOS Netflow – A Technical Overview", last accessed: February 2006, http://www.cisco.com/en/US/products/ps6601/, last visited 01/08/2006.

A Protocol for Atomic Deployment of Management Policies in QoS-Enabled Networks

Rodrigo Sanger Alves, Lisandro Zambenedetti Granville,
Maria Janilce Bosquiroli Almeida, and Liane Margarida Rockenbach Tarouco

Federal University of Rio Grande do Sul (UFRGS) - Institute of Informatics
Av. Bento Gonçalves, 9500 - Bloco IV - Porto Alegre, RS - Brazil
{sanger, granville, janilce, liane}@inf.ufrgs.br

Abstract. This paper presents a novel protocol to support the atomic deployment of management policies for networks with quality of service (QoS) support. The necessity of such a protocol comes from the fact that faulty policy deployments lead to situations where the required QoS is not provided to network users but still consumes network resources such as bandwidth. In addition to the protocol definition, we present a Web services-based implementation and an analysis of the proposed protocol in a policy-based architecture for the management of differentiated services (DiffServ)-enabled networks.

1 Introduction

Policy-Based Network Management (PBNM) is an effective approach to promote the management of networks with quality of service (QoS) support [1]. Although several PBNM architectures have been proposed in the recent years [2], the IETF (Internet Engineering Task Force) policy architecture is probably the one most widely recognized. In fact, such architecture is not formally defined in the IETF documents, but it seems to be a wide acceptance on a common set of architectural elements with well recognized roles. A **policy tool** is used by the human network operator to define QoS policies that are stored in a **policy repository** for future reuse and/or deployment. **Policy Decision Points** (PDPs) distributed along the managed network are responsible to receive those policies and translate them to QoS-related configuration actions applied to **Policy Enforcement Points** (PEPs), which are found inside the managed network devices.

PDPs and PEPs communication is critical to provide a feasible policy deployment. The IETF has been actively working on this area defining PDP/PEP communication protocols such as COPS (Common Open Policy Service) and COPS-PR (COPS Usage for Policy Provisioning) [3]. In addition, the IETF has also been working on the definition of policy information models such as PCIMe (Policy Core Information Model extensions) [4] and QPIM (Quality of Service Policy Information Model) [5].

Although policy information models and PDP/PEP communication are areas whose developments are clearly perceived, less development is carried out concerning the communication between the policy tool and PDPs. For example, there is no standardized or under standardization protocol to cover such communication. The lack of standardization leads to a scenario where each PBNM system defines and implements its own

G. Parr, D. Malone, and M. Ó Foghlú (Eds.): IPOM 2006, LNCS 4268, pp. 132–143, 2006.

policy tool/PDP protocols. Such protocols can be error-prone, which enables the presence of faulty situations in the policy deployment process.

In this paper we are particularly interested in investigating the requirements and implementation of an atomic policy deployment, which addresses the ability of a PBNM system to ensure that a policy is deployed on all target PEPs, or it is not deployed at all, i.e., if one single target PEP fails to have a policy deployed, then all other PEPs where the same policy has been already deployed must rollback to the state prior to the policy deployment.

The remainder of this paper is organized as follows. Section 2 states the problem investigated in this paper presenting a testing, initially faulty scenario. Section 3 introduces the proposed approach and associated protocol. Section 4 shows the proposed protocol implementation using the Web Services technology. The protocol network usage is analysed in section 5. Section 6 presents related work, and finally Section 7 concludes this paper with final remarks and future directions.

2   Faulty Policy Deployment Process

Policy deployment involves, considering the IETF PBNM architecture, at least two phases of communication. First, the network tool must contact the appropriate PDPs in order to transfer a policy from the policy repository. On a second phase, each PDP interacts with associated PEPs to perform the policy deployment. In this last step, it is possible to use two approaches for PDP/PEP interaction: top-down (or provisioning), where PDP initiates policy deployment contacting PEPs; and bottom-up (outsourcing), where PEPs contact PDPs in order to determine, for example, whether a new traffic can be admitted or not. For simplicity, we will consider only the deployment of policies through the provisioning approach because it is more adequate for DiffServ networks.

2.1   Policy Deployment

Once a policy is transferred to a PDP, the PDP needs to evaluate the policy to determine if its condition clause evolves to true. When the policy conditions evolve to true the PDP should translate the policy action clause to configuration actions on the target PEPs. At this point, the policy may fail due to the limited resources of a target PEP. For instance, if a policy action states to allocate 6 mbps of bandwidth but the available link is a backup one limited to only 2 mbps, then the policy deployment fails.

Recovering this policy deployment can be accomplished by different approaches. However, the point we are investigating in this paper is the fact that, in some situations, the failure of a single policy deployment on a PEP, regardless the recovery approach used, not only affects the policy on that PEP but also on all the other PEPs where the same policy should be deployed.

2.2   Distributed Policy Deployment Failure

In QoS management, policy deployment is essentially a distributed procedure. For example, let's consider the test scenario presented in Figure 1, where two PDPs receive a policy to be deployed on four routers, but at router R2 the deployment fails.

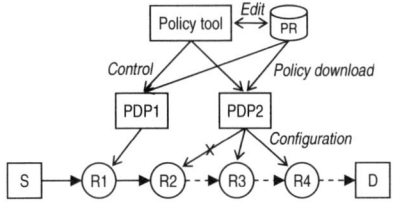

Fig. 1. Policy deployment failure

In order to allocate network bandwidth, the policy tool needs to transfer (via pull or push) the appropriate policy from the policy repository (PR) to the PDPs that control the PEPs inside the network routers (R1, R2, R3, and R4) between the source device (S) and the destination device (D). For all PEPs inside devices where the policy deployment succeeds (R1, R3, and R4) the required bandwidth will be allocated. However, if a single policy deployment fails (R2), the remainder PEPs will still provide the allocated bandwidth but that is now useless since one device with no bandwidth allocated is enough to compromise to whole path. It means that even though routers R3 and R4 provide the bandwidth requested in the transferred policy, the traffic sent from S to D will be compromised by router R2, where the policy deployment has failed.

2.3   Policy Deployment Rollback

Once a single policy deployment fails on the previous scenario, one should ensure that the other successful deployments rollback in order to release the network bandwidth allocated on the successful PEPs. Otherwise, the bandwidth would be wasted serving a traffic that will probably not use all allocated bandwidth due to the failure on a single PEP. Rolling back a policy deployment involves, through the policy tool, the explicitly indication that all PDPs must locally rollback too. The most important point here is to notice that currently this distributed rollback procedure is only accomplished through the manual intervention of the network operator.

In this scenario we envisage that for some applications (e.g., QoS management), policy deployment must work for all target PEPs, or for none target PEPs at all, i.e., policy deployment should be an atomic distributed operation. One of the difficulties in defining such an atomic deployment is that the communication between the policy tool and PDPs is not standardized, as presented in the introduction section. To work around this situation, in the next section we present a new solution and associated protocol used in the communication between the policy tool and PDPs.

3   A Protocol for Atomic Policy Deployment

A network operator that wants to deploy a policy on the network must first select a policy p from the policy repository. Also, he or she needs to pickup, in the policy tool, those PEPs where the selected policy will be deployed (*tPeps*). The policy tool must then provide a key (*delpId*) that identifies the deployment of p on *tPeps*. Given the PEPs from *tPeps* the policy tool must also compute the target PDPs (*tPdps*) that need to be contacted in order to proceed with the policy deployment. Summarizing, we have:

- *p* is the policy to be deployed
- *tPeps*={pep1, pep2...pep*n*} is the set of target PEPs
- *tPdps*={pdp1, pdp2...pdp*m*} is the set of target PDPs
- *deplId* is the deployment identification of *p* on *tPeps*

After having *p*, *tPeps*, *deplId*, *tPdps*, and the associations between PDPs and PEPs all defined, the policy tool contacts the PDPs from *tPpds* and transfers to each of them: (a) the policy *p*, (b) the deployment identification *deplId*, and (c) the PEPs where *p* should be deployed.

3.1 PDPs and Policy State

A policy *p* inside a PDP has a policy state: *invalid, inactive, active*. In order to present these different states, let's take the sample policy from Figure 2.

```
if (day >= 1/2/2006) and (day < 12/24/2006) and
   ((dWeek=Friday) or (dMounth=Last)) and
   (time >= 8pm) and (time <= 9pm) and
   (sourcePort = 80)
then bandwidth = 500 kbps
```

Fig. 2. Policy example

This policy definition declares that January 2, 2006 is its starting time. Probably the network operator, however, will order the policy tool to transfer such policy to target PDPs before January 2. In this case, just after being transferred to a PDP, the policy state is *invalid* because its starting time has not been reached yet. The sample policy also states that the ending time is December 24, 2006. After this time, the policy is *invalid* too, and should be removed from PDPs.

When the starting time is reached, the policy state changes to *inactive*, i.e., the policy is now valid but it is not active yet. Once a policy is inactive, it needs to evolve to *active* in order to effectively contact the target PEPs controlled by a PDP to configure them. In our example, the sample policy evolves to *active* - and then allocates 500 kbps of bandwidth for downstream HTTP traffic (using port 80) on associated PEPs - only on Fridays and on the last day of months at 8 pm. At 9 pm the policy will not be active anymore, going back to *inactive*. Figure 3 summarizes the common sequence of state changes of a policy inside a PDP.

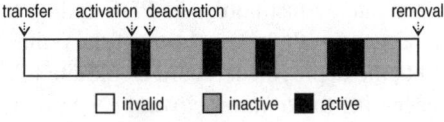

Fig. 3. Policy states inside a PDP

3.2   Policy Deployment Coordinator

Atomic deployment is accomplished through atomic activation. It means that each time a policy *p* is activated inside a PDP, it must be also in all other PDPs from *tPdps* in order to have all PEPs from *tPeps* properly configured. Also, each time a policy is deactivated, it must be deactivated in all *tPdps* as well. We introduce a new element in the PBNM architecture named Policy Deployment Coordinator (PDC), whose responsibility is to orchestrate atomic policy activations/deactivations. That is possible through full-duplex communications among the PDC and the PDPs on *tPdps*.

PDC coordinates the activations of a policy *p* using a data structure whose entries are indexed by *deplId*. Such structure, named *deployment record*, includes an integer that indicates the current state of a deployment, and a vector of PDPs that stores the policy activation state for each involved PDP. Computing such vector allows the PDC to identify the global activation state of a policy. In summary:

- DeplKey: the unique identification of a policy deployment (*deplId*)
- DeplState: [0:inactive 1:activating 2:activated]
- DeplVector PDPs: the set of PDP for this policy deployment (*tPdps*)
- DeplVector state: the state of each PDP ([0:unknown 1:success 2:failure])

Although we have defined the PDC as a new element in the PBNM architecture, it can be physically implemented together in the policy tool or even internally on a PDP.

3.3   Policy Transfer and Activation Operations

Once the network operator has selected the policy to deploy and the PEPs where such policy should be applied, i.e., the operator has defined *p* and *tPeps*, the policy tool must compute *deplId* and *tPdps*. Then the PDC *RegisterPolicyDeployment* operation must be called by the policy tool in order to create a deployment record of *p* on PDC. The operation receives as input arguments *deplId* to be associated with DeplKey, and *tPdps* to instantiate the DeplVector PDPs. The deployment state DeplState is initialized with *inactive* (0), and each element of DeplVector state is initialized with *unknown* (0).

After registering the policy deployment at the PDC, the policy tool transfers (via pull or push) the policy *p* to all PDPs on *tPdps* using a *PolicyTransfer* operation exposed by each PDP. This operation expects as input argument the policy *p*, the list of PEPs this PDP should configure (which is a subset of *tPeps*), and *deplId*. After that, all policy deployment information will have been transferred.

Internally, each PDP must check the temporal constraints of *p* in order to activate such policy when required. Since PDPs' clocks are not necessarily fully synchronized, the activation of policy *p* can be triggered on slightly different moments in different PDPs. The first PDP to activate *p* must notify the PDC calling the PDC *InitiateActivation* operation, and then proceed with the execution of the policy actions.

Once notified, PDC sets the deployment record DeplState to *activating* (1) and back notifies all other PDPs accessing the *InitiateActivation* operation exposed by them. This notification orders the remaining PDPs to immediately initiate the activation of *p*. PDPs that activate a policy due to a notification via *InitiateActivation* does not need to notify the PDC anymore, since the PDC will ignore all other calls to its *InitiateActivation* operation (Figure 4).

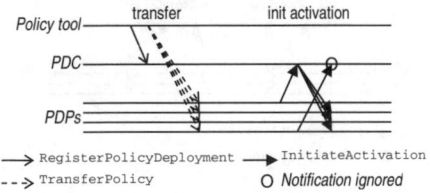

---> RegisterPolicyDeployment ——▶ InitiateActivation
--→ TransferPolicy   O Notification ignored

Fig. 4. Policy transfer and activation

3.4 Policy Activation Commit and Rollback

After policy activation has been initiated, all PDPs are configuring the appropriate PEPs. At this moment, fails can happen, and the PDC needs to be notified on this. While in the *activating* state, the PDC waits for reports on the activation results from PDPs.

As soon as a PDP configures all associated PEPs it should call the PDC *Activation-Success* operation. However, as soon as the PDP realizes that the configuration of any single PEP failed, it should rollback the configuration on all associated PEPs and notify the fail calling the PDC *ActivationFail* operation.

For each *ActivationSuccess* call, the PDC sets the state of the caller PDP in DeplVector to *success* (1). When all positions of DeplVector status are set to 1, the PDC calls back a *Commit* operation of each PDP, then indicating that the whole activation process worked with no fails (Figure 5).

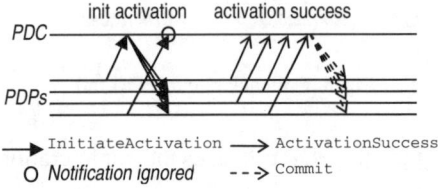

——▶ InitiateActivation ——> ActivationSuccess
O Notification ignored   --→ Commit

Fig. 5. Activation initialization and commit

If one PDP notifies the PDC through *ActivationFail*, then the PDC immediately notifies all other PDPs, except the caller, calling the *Rollback* operation (Figure 6). Each PDP, on its turn, contacts the associated PEPs to remove the configuration previously deployed. Again, how PDP and PEPs interact with each other to effectively rollback is another implementation detail not covered by our protocol.

Since some fails can happen not only on PEPs, but also on PDPs and PDC, some timeouts have been additionally defined. The first one is located at the PDC. It waits for activation reports from PDPs until a maximum time. If no more PDPs report until this maximum time, and there are some DeplVector status still *unknown* (0), the PDC assumes that the lack of notifications is due to faulty PDPs and then the *Rollback* notification is issued on all PDPs (Figure 7).

The second timeout is located at PDPs. After calling the PDC *ActivationSuccess* operation, the PDP waits for a *Commit* or *Rollback* call. If none of them is called in a

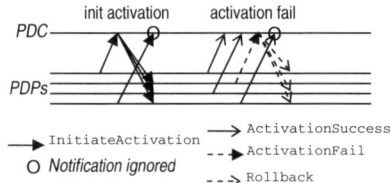

Fig. 6. Activation initialization and rollback

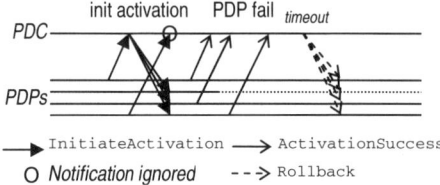

Fig. 7. Rollback due to PDP fail

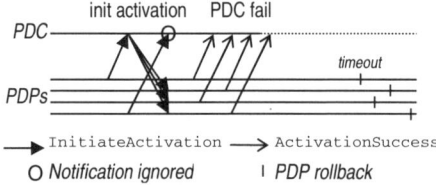

Fig. 8. Rollback due to PDC fail

limited amount of time, then the PDP rollbacks the configuration by default, assuming that the PDC is faulty (Figure 8).

3.5   Policy Deactivation and Removal

Deactivating a policy is a straight forward operation. The first PDP that detects that the active policy should be deactivated must notify the PDC calling the *Deactive* operation. The PDC will forward this notification to all remaining PDPs calling the *Deactive* operation exposed by PDPs. In addition, the PDP sets the DeplStatus to *inactive* (0) and resets all positions of the DeplVector status *unknown* (0) in order to be ready to coordinate next policy activations.

Although a policy is deactivated from time to time, it remains valid inside PDPs until the policy ending time is reached, when it finally evolves to *invalid* again. In this case, the first PDP that perceives that a policy is *invalid* notifies the PDC through the *RemovePolicy* operation, which is quite similar to the *Deactive* operation (Figure 9) except that it will remove all instances of the policy from all associated PDPs, and frees the memory used to store the policy deployment record at the PDC.

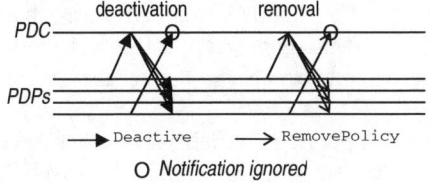

Fig. 9. Policy deactivation and removal

4   Protocol Implementation

We have implemented the proposed protocol in a prototype system called QAME (QoS-Aware Management Environment) [6]. QAME already supports policy based management of DiffServ networks. An LDAP [7] repository is used to store QoS management policies defined through graphical user interfaces and wizards. We have extended the system PDPs to include the atomic deployment support defined before, as well as have integrated a new PDC into the existing policy tool.

Figure 10 left presents the final architecture of our prototype system. It is important to highlight again that the PDC has been implemented together with the policy tool.

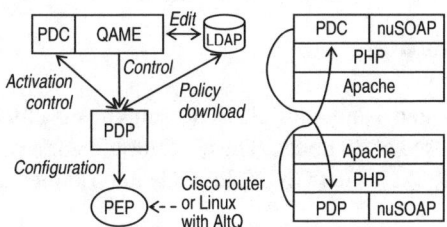

Fig. 10. PBNM architecture with PDC

Policies are stored in an LDAP directory based on PCIMe [4] and QPIM [5], as recommended by IETF. The policy tool transfers policies to a set of PDPs (although just one is presented in the figure) using the pull approach, i.e., each PDP retrieves policies from the LDAP directory.

Currently, there are two different PDPs to support the proper configuration of two different PEPs: Cisco DiffServ routers and Linux routers with ALTQ [8]. PDP commits and rollbacks in Cisco routers are supported manipulating router tentative internal configurations that may either be discarded (roolback) or evolve to permanent configurations (commit). In the case of ALTQ, the associated PDP backs up the target PEP configuration via SSH before applying a new configuration. If a rollback is required, then the backup configuration is deployed again.

Figure 10 right presents a simplified view of the internal architecture of both PDC and PDP. Web Services support is based on the Apache HTTP server, PHP, and nu-SOAP library. The PDC itself is a PHP script that implements the Web Service that exposes the PDC operations. The PDP is also a PHP script that exposes the PDP Web Services operations, but it is also composed of a Linux daemon that constantly evaluates the deployed policies in order to check whether they must be translated to a PEP configuration.

The proposed protocol is supported through a set of Web Services operations implemented in the elements of the PBNM architecture. Using Web Services as the basis for our protocol is interesting due to the following points:

- Web Services traffic is usually carried by SOAP (Simple Object Access Protocol) [9] over HTTP messages normally accepted by firewalls. It allows an easier PDP deployment along different administrative domains;
- Since HTTP runs over TCP - which is a connection-oriented reliable transport protocol - implementing our protocol via Web Services is interesting because they provide a reliable delivery of protocol messages;
- Web Services RPC is easier and more generic than defining protocol PDUs and data encodings.

We have implemented our protocol using the nuSOAP [10] Web Services API for PHP. This API is largely used by the free software community of Web Services developers and is relatively fast if compared with other Web Services PHP APIs.

5   Protocol Evaluation

The protocol was evaluated using multiple PDPs running on a high performance cluster named LabTeC, composed of 20 nodes. The configuration of each cluster node is: Dual Pentium III 1.2 Ghz, RAM 1 GB, GNU/Linux Gentoo (kernel 2.6.14), Apache 2.0.54 and PHP 4.4.0.

In our evaluation we have used an increasing number of PDPs: 2, 3, 4, 6, 8, 12, 16, 24, and 32. Each PDP runs in a dedicated cluster node except for 24 and 32 PDPs, where some nodes have run more than one PDP internally. A last additional cluster node has been used to host the PDC. The performance results have been obtained running the TCPDump software to capture packets in the node running the PDC.

The difference in the internal clock of each PDP affects the number of InitiateActivation messages sent from PDPs to the PDC. If such difference is minimal, more messages will be sent simultaneously. On the other hand, the greater the difference, less messages will be sent. Therefore, in order to have results closer to the worst case, we have adjusted the clocks of each node to be as close as possible. To achieve this approximated clock synchronization we have used the Network Time Protocol (NTP).

During the tests no actual device configuration has been performed. In fact, each PDP at activation or deactivation time executed no action except sending the "ActivationSuccess" or "ActivationFail" messages. Such implementation was adopted to simplify the test scenario requirements, since the evaluation goal is focused only in the protocol overhead.

The network usage and execution time obtained from the activation tests are presented in Figure 11.

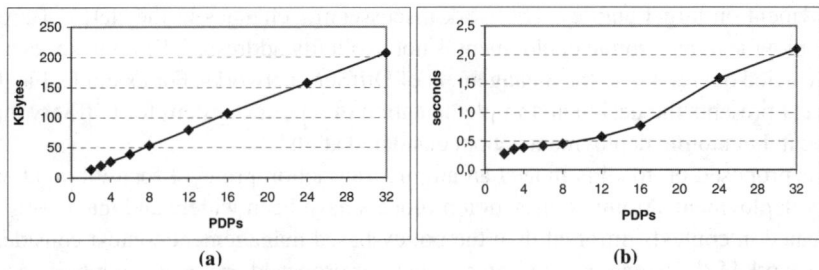

Fig. 11. (a) Network usage and (b) execution time for activation tests

As expected, the network usage is directly proportional to the number of PDPs. This behavior can be explained by the fact that the number of messages exchanged by the protocol during an activation has almost no variation except for the InitiateActivation messages sent by PDPs. It means that the network usage can be reasonably calculated solely based on the number of PDPs.

The execution time seems to be scalable until 16 PDPs, but for 24 and 32 PDPs the execution time grows up more quickly. The reasons is that up to 16 PDPs each cluster node has hosted a single PDP. In the tests with 24 and 32 PDPs some nodes have hosted more than one single PDP. This intra-node resource concurrency has thus determined this time difference.

The network usage and execution time obtained from the deactivation tests are presented in Figure 12.

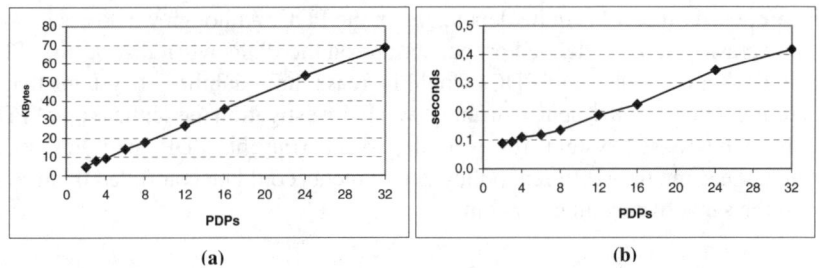

Fig. 12. (a) Network usage and (b) execution time for deactivation tests

The results obtained in the deactivation tests are similar to those obtained in the activation tests. It is important to notice, however, that the execution time has suffered a minor influence of the utilization of more than one PDP per cluster node. This is due to the fact that the deactivation process requires less computational resources from the host node than the activation process.

6   Related Work

Dulay et al. [11] presents a model for the deployment of policies that are defined using the Ponder policy language. Deployment objects are used to coordinate the policy deployment on target entities. The work discusses the changes in the states of a policy deployment, but atomic deployment is not explicitly addressed. Several works have been carried out related to the management of DiffServ networks. For example, Flegkas et al. [1] presents the design and implementation of a policy system for DiffServ management, but atomic deployment is not considered at all.

Our proposed protocol is indeed an atomic transaction protocol focused on atomic policy deployment. Atomic transaction protocols have been widely and intensively investigated in contexts different than the policy-based management context considered in this work [12]. In general, one can say that our proposed protocol is a modified version of the 2PC (two-phase commit) protocol [13] adapted to policy-based management environments and necessities.

7   Conclusions and Future Work

In this paper we have introduced a novel protocol for atomic deployment of policies. Such atomic deployment is specially important for the management of QoS-enabled networks, where, for instance, configuration actions to allocate bandwidth should work in all target network elements.

We have argued that implementing the proposed protocol using the Web Services technology is an interesting option because Web Services are widely supported by development languages and platforms, and because it is easier to define a protocol as a set of operations instead of defining new PDUs and data encoding rules. In addition, the analysis of the implementation showed that, for the current networks, the traffic generated by PDC and PDPs is low and can normally fit in modern networks.

The proposed protocol is quite dependent on the PDC. Although we do assume that PDC may crash, and have defined proper actions on the PDPs when that occurs (PDPs will rollback), a redundance of PDCs would increase the reliability of systems that require atomic policy deployment. Currently we are investigation the replication of PDCs along the policy-based system, for example, considering that PDPs may host PDCs, and that at the same time different policy deployments could be controlled by different PDCs of the same management system.

References

1. Flegkas, P., Trimintzios, P., Pavlou, G.: A Policy-based Quality of Service Management System for IP Differentiated Services Networks. IEEE Network **16**(2) (2002) 50–56 issn: 0890-8044.
2. Guo, X., Yang, K., Galis, A., Cheng, X., Yang, B., Liu, D.: A policy-based network management system for IP VPN. In: ICCT. Volume 2. (2003) 1630–1633
3. Chan, K., Seligson, J., Durham, D., Gai, S., McCloghrie, K., Herzog, S., Reichmeyer, F., Yavatkar, R., Smith, A.: RFC 3084: COPS Usage for Policy Provisioning (COPS-PR). (2001) Status: PROPOSED STANDARD.
4. Moore, B.: RFC 3460: Policy Core Information Model (PCIM) Extensions. (2003) Status: PROPOSED STANDARD.
5. Snir, Y., Ramberg, Y., Strassner, J., Cohen, R., Moore, B.: RFC 3644: Policy Quality of Service (QoS) Information Model. (2003) Status: PROPOSED STANDARD.

6. Granville, L., Ceccon, M., Tarouco, L., Almeida, M., Carissimi, A.: An Approach for Integrated Management of Networks with Quality of Service Support Using QAME. In Festor, O., Pras, A., eds.: IEEE/IFIP 12th International Workshop on Distributed Systems: Operations and Management, DSOM 2001, Nancy, France, October 15-17, 2001. Proceedings. (2001) 167–178 isbn: 2-7261-1190-4.
7. Hodges, J., Morgan, R.: RFC 3377: Lightweight Directory Access Protocol (v3): Technical Specification. (2002) Status: PROPOSED STANDARD.
8. Cho, K.: Managing Traffic with ALTQ. In: FREENIX Track: USENIX Annual Technical Conference. (1999) 121–128
9. Mitra, N.: SOAP Version 1.2 Part 0: Primer. (2003) W3C Recommendation 24.
10. Ayala, D.:   NuSOAP - Web Services Toolkit for PHP. (2006) Available at http://dietrich.ganx4.com/nusoap/.
11. Dulay, N., Lupu, E., Sloman, M., Damianou, N.: A Policy Deployment Model for the Ponder Language. In: IEEE/IFIP International Symposium on Integrated Network Management (IM'2001), IEEE Press (2001) 529–543
12. Tang, L.: Verifiable transaction atomicity for electronic payment protocols. In: 16th International Conference on Distributed Computing Systems (ICDCS'96), IEEE Computer Society (1996) 261
13. Lampson, B.W., Lomet, D.B.: A new presumed commit optimization for two phase commit. In: Proceedings of the 19th International Conference on Very Large Data Bases, Morgan Kaufmann Publishers Inc. (1993) 630–640

Towards Autonomic Network Management for Mobile IPv4 Based Wireless Networks*

Dong-Hee Kwon, Woo-Jae Kim, Young-Joo Suh, and James W. Hong

Departments of Computer Science and Engineering
Pohang University of Science and Technology (POSTECH)
{ddal, hades15, yjsuh, jwkhong}@postech.ac.kr

Abstract. The rapid progress of wireless communication technologies has opened a possibility to offer various types of communications to users irrespective of their locations. The all-IP based wireless networks have been proposed and the Mobile IP protocol is considered as one of prominent candidate frameworks to support a seamless mobility of users. However, to our best knowledge, there are few research efforts to design and develop a network management system targeted for Mobile IP based wireless networks. In this paper, we introduce the concept of *autonomic wireless network management* which utilizes SNMP agents to manage more intelligently through the self-management functionality. We present a design of autonomic wireless network management system (AWNMS) and its prototype implementation. The currently implemented system is fully compliant with Mobile IP MIB and provides management functions such as network topology auto-discovery, mobility tracking function, etc. The implemented system is validated and examined in a Wireless LAN based test-bed network.

1 Introduction

The rapid progress of wireless data communication technologies ranging from Wireless Local Area Networks (WLANs; e.g., IEEE 802.11[1]) to Wireless Metropolitan Area Networks (WMANs; e.g., IEEE 802.16 [2]) and third-generation cellular systems [3] have opened a possibility to offer various types of communications to users irrespective of their locations. This development tendency of wireless network technologies has accelerated the diversity of wireless network devices and allowed user devices to move freely within large geographical areas and changed the form of handover management. The handover management in wireless networks such as cellular networks is carried out by technology specific mechanisms since only the intra-technology handovers are involved given the same type of wireless network and single mode terminals. However, the need for integrating different types of wireless networks and the emergence of multi-mode enabled mobile

* This research was supported by the MIC (Ministry of Information and Communication), Korea, under the ITRC (Information Technology Research Center) support program supervised by the IITA (Institute of Information Technology Assessment)" (IITA-2005- C1090-0501-0018) and by the Electrical and Computer Engineering Division at POSTECH under the BK21 program of the Ministry of Education, Korea.

G. Parr, D. Malone, and M. Ó Foghlú (Eds.): IPOM 2006, LNCS 4268, pp. 144–155, 2006.

terminals have pushed the traditional handover functionality at L2 layer to the generic IP layer that serves the rendezvous point of underlying wireless technologies. Accordingly, the need for migration of technology specific core infrastructures to all-IP based networks has been identified and great efforts are placed in this direction. Mobile IPv4 [4] and Mobile IPv6 [5] which have been standardized by Internet Engineering Task Force (IETF) are the results of such efforts.

In general, a network management system provides useful functionalities and information to network administrators for efficiently managing target networks. Simple Network Management Protocol (SNMP) [6] is one of the most widely used protocols for IP based network management. SNMP enables network administrators to manage network performance, find and solve network problems, and plan for network growth. The management functional areas that should be provided by network management systems are fault, accounting, configuration, performance, and security [7]. The Mobile IP based wireless networks also need a network management system to efficiently manage and use networks. However, to our best knowledge, there are few research efforts to design and develop a network management system on mobile wireless networks, typically focusing on the Mobile IPv4 protocol. In this paper, we design and implement a network management system for the Mobile IPv4 based wireless networks.

We first introduce the concept of *autonomic wireless network management* which utilizes SNMP agents to manage more intelligently through the self-management functionality. The SNMP agents in our autonomic wireless network management system (AWNMS) exchange network information such as network resource status and security information with neighbor agents. In the mobile wireless network, the seamless mobility is important to users for continuous communication during handovers. IETF has standardized new protocols such as the Candidate Access Router Discovery (CARD) [9] for supporting the seamless mobility. However, in the proposed AWNMS, we integrate these functionalities to SNMP agents using the self-management function and do not require such new protocols for seamless mobility.

In this paper, we also present results of the prototype implementation based on our Mobile IPv4 protocol stack. The currently implemented system is fully compliant with Mobile IP MIB [8] and provides management functions such as network topology auto-discovery, mobility tracking function, etc. To support these management functions, our AWNMS uses the integrated information from multiple managed objects for providing advanced functions in the wireless networks. The implemented system is validated and examined in a Wireless LAN based test-bed network.

2 Related Work

Mobile IPv4/IPv6 [4, 5] proposed by IETF provides a basic node mobility management scheme in IP networks. When a mobile node (MN) moves from one subnet to another ((1) in Fig. 1), it requests its home agent (HA) to update its binding to receive continuous service ((2) binding update in Fig. 1). In case of Mobile IPv6, configuration of new care-of-address (CoA), which denotes the current location of the MN, is required prior to sending a Binding Update message to HA. A binding maintained by the HA is an association of a MN's home address and its CoA. When the HA has the binding for the MN, the HA intercepts any packets destined to the MN, and tunnels them to the MN's CoA ((3) and (4) in Fig. 1). Thus, it is necessary for the MN to register its current point of attachment to the HA whenever it handovers to another network. Thus, the Mobile IP scheme enables the MN to move freely in the Internet without any disruption of data service.

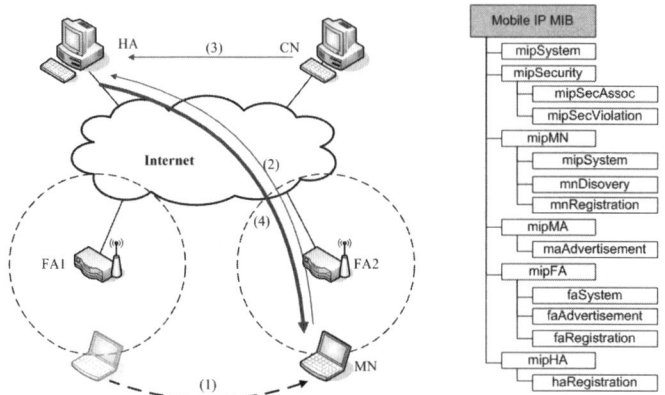

Fig. 1. Mobile IP operation   **Fig. 2.** Mobile IP MIB

To achieve this, three entities (HA, FA, and MN) defined in Mobile IPv4 should interact with each other by exchanging signaling messages to update and synchronize important information such as the state of each entity and the binding information. To endow a network management system to control and monitor the protocol operation of Mobile IPv4, the definitions of Managed Objects for IP Mobility Support using SMIv2 (Mobile IP MIB) has been standardized [8].

Table 1. Relationships between Mobile IP MIB groups and entities

	MN	FA	HA
mipSystem	O	O	O
mipSecurity	O	O	O
mipMN	O		
mipMA		O	O
mipFA		O	
mipHA			O

The Mobile IP MIB, shown in Fig. 2, consists of six main groups which are related to the HA, FA, and MN. The relationships between six groups and three entities are shown in Table 1. First of all, mipSystem and mipSecurity groups represent the common characteristics to all entities of Mobile IP. The mipSystem group shows the UP/DOWN status and supported packet encapsulation type of each entity, where the mipSecurity group represents the security related information such as security algorithm, security mode, security violation of SPI, etc., exchanged among entities during the registration of Mobile IP. The mipMA group is the common MIB used by two types of Mobility Agents (FA and HA), and shows agent specific information such as agent advertisement. The remaining Mobile IP MIB groups (mipMN, mipFA, and mipHA group) provide the entity-specific information.

3 Autonomic Network Management for Mobile IPv4 Based Wireless Networks

In this section, we introduce the concept autonomic wireless network management and architecture for Autonomic Wireless Network Management System (AWNMS), which makes SNMP agents to manage wireless networks more intelligently through the self-management functionality.

3.1 Autonomic Wireless Network Management

In wireless networks, it is always preferable that mobile nodes are supported the seamless mobility for continuous communication regardless of their handover events. The seamless mobility means that mobile nodes experience the minimized handover latency and can communicate continuously with the communicating partner moving from one network to the new network. For the seamless mobility, network entities in Mobile IPv4 should have active functionalities such as dynamic and adaptive resource management, security negotiation, etc. For example, if the application used in the mobile node requires certain level of QoS for successful communication, network entities should carefully accept the handover mobile node. When the network entity decides that the remaining resource is enough to support the handover mobile node, it can accept the handover request from the previous network entity or the mobile node. Otherwise, the network entity should reject the handover request. Also, if the mobile node initiates the QoS or security negotiation after connecting to the new network, it cannot communicate with its communicating partner until the negotiation procedure is completed. This results in long handover latency.

As the Mobile IPv4 only specifies a basic mobility management scheme in IP networks, several new protocols and schemes have been proposed to overcome the drawback of Mobile IPv4. One of them is the proactive QoS and security negotiations before the actual handover operation. For this, some protocols for information exchange between Mobility Agents or Access Points, such as CARD [9] and Inter-Access Point Protocol [14], are proposed. We have integrated these functionalities to the SNMP management framework using the autonomic computing concept [15]. The autonomic management in this paper means that a management system can adaptively monitor, analyze and control its managed systems without any intervention of network managers or operators. In this paper, we address that the functionality of a SNMP agent can be easily extended to have autonomic management functionality and thus a SNMP management framework can provide required operations for the seamless mobility. In the proposed system, we define this component in the active SNMP agent to support this autonomic management functionality.

The network entities such as Mobility Agents also should have adaptive resource management functionality to efficiently manage their radio resources. When a high priority user requests more resources to the Mobility Agent, the Mobility Agent may examine the radio resource usage of each user and allocate resources to the high priority user by preemptively reducing the allocated resources to the low priority users. This resource management functionality is invoked when the user coming into the new network has higher priority than the existing users and the remaining resource cannot support the user's request. This operation also can be integrated to the SNMP management framework using the autonomic network management concept; the self-management in the proposed system. In the proposed system, the pre-defined thresholds and the radio resource usages are stored and retrieved to/from the integrated information repository. The active SNMP agent periodically monitors and updates this information based on the basic SNMP operation. If the active SNMP agent exchanges this information with neighbor agents, it may recommend the mobile node to handover to the neighbor network in order to distribute the network load.

In summary, the autonomic concept of the active SNMP agent has two components; the self-management and the manager function which is the ability to exchange information with neighbor agents.

3.2 Architecture of the Autonomic Wireless Network Management System

Fig. 3 illustrates the architecture of AWNMS, where SNMP agents have the manager functionality to autonomously manage target systems and networks through a self-management scheme. The self-management means that a SNMP agent, which is referred to as an 'active SNMP agent' in this paper, can dynamically configure managed systems without the manager or the network operator.

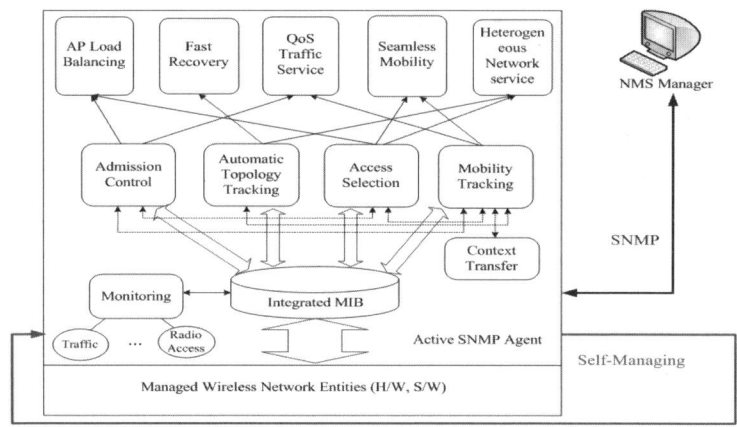

Fig. 3. The architecture of AWNMS

The Integrated MIB (shown in Fig. 3) is a logical management information repository which is maintained by an active SNMP agent through gathering and combining individual information of pre-defined MIBs such as Mobile IP MIB and MIB-II. Therefore, the Integrated MIB operates as the meta-MIB and the basis of the other components. The managed objects of the Integrated MIB, whose example instance is illustrated in Table 2, provide more useful and intuitive information to users than the basic MIBs. As shown in Table 2, when the manager needs the network load information, it just retrieves the information of FALoad managed object through the active SNMP agent. But, without this integrated MIB, we may have to make the

Table 2. An example instance of managed objects of an Integrated MIB

Management Object	Information Sources	Description
MNLocation	mnFATable and mnRegistrationTable in Mobile IP MIB	Movement history of each MN. It is maintained as table structure.
FALoad	faVisitorTable in Mobile IP MIB and ifTable in MIB-II	Network load information in each FA. It counts the number of bytes destined to each mobile node.
NeighborAgent	ipRouteTable in MIB-II and newly defined topology information	Neighbor mobility agents information of each mobility agent.
TunneledTraffic	haMobilityBindingTable in Mobile IP MIB and ifTable in MIB-II	Number of transmitted bytes to each MN using tunneling.

manager to gather separately each information and process them, and this makes implementing the SNMP manger even harder and more complex.

In wireless networks, information exchange such as context transfer between network elements is required for efficient mobility management [9, 10]. As the manager can exchange information with other managers, the manager function of the active SNMP agent also can exchange mobility related information, such as context and security information, with neighbor agents having the manager function (Context Transfer component in Fig. 3). The Context Transfer component also can exchange the network capability information with each other.

Based on the Integrated MIB, the active SNMP agent provides four basic components as shown in Fig. 3; admission control, automatic topology tracking, access selection, and mobility tracking. Each component provides useful information to the manager function of the active SNMP agent. As a general admission control function for wireless networks, the Admission Control component provides the network status information such as currently used radio resources and required resources to support new communications. The Integrated MIB can have the resource utilization information using the monitoring component. The Automatic Topology Tracking component provides the network connectivity information. When we consider the node mobility in the Mobile IPv4 network, the network topology discovery functionality becomes more important in the wireless network management system. This information can be retrieved through the general MIB (IP group in MIB-II) and Mobile IP MIB (MN/HA/FA registration group in Mobile IP MIB). The Access Selection component provides the network capability information of neighbor networks. This component helps the network discovery and evaluation of the target network to which the mobile node will handover operations required by the handover operation and provides base information to determine the handover is triggered by network-based decision or by node based decision. Finally, the Mobility Tracking component provides the mobility history and pattern information of mobile nodes. The mobility history information is stored in the Integrated MIB through comparing the care-of address information of the mobile node in the Mobile IP MIB. Using the mobility pattern of each mobile node, the active SNMP agent can exchange the QoS and security information prior to the actual handover for fast negotiation and reduction of the handover latency.

The manager function of the active SNMP agent provides five basic functions based on above components as shown in Fig. 3; AP load balancing, fast recovery, QoS traffic service, seamless mobility, and heterogeneous network service. The AP load balancing and QoS traffic service focus on the communication quality of mobile nodes, and fast recovery, seamless mobility and heterogeneous network service focus on the node mobility. In wireless networks managed by the AWNMS, each AP can maintain its own measurement information such as the number of servicing mobile nodes, maximum capacities and the measure of load, and the neighbor APs exchange the maintained information with each other. Based on this collected information, the AWNMS conducts AP load balancing and QoS traffic service. In AWNMS, the active SNMP agents on APs regulate the load of each AP using the admission control and access control components. The active SNMP agents also maintain user QoS profile and application QoS profile in the Integrated MIB, and adjust QoS parameters or decide to switch network access for better QoS using the mobility tracking and admission control components.

In the Mobile IPv4 based wireless networks, the node mobility should be supported and transparent to users. In AWNMS, the active SNMP agent utilizes the node's mobility pattern to support a seamless, transparent handover using the mobility tracking component. The active SNMP agent can predict the node movement based on the mobility pattern, and makes the environments to communicate immediately through the negotiation with node's profile and QoS profile. The negotiation of node profiles can be conducted between different radio access networks based on the automatic topology tracking and access selection components. Therefore, the AWNMS can support the heterogeneous network service and fast recovery as well as the seamless mobility.

4 Implementation

In this section, we present the Mobile IP MIB and Integrated MIB implementation, and the implementation details of the prototype AWNMS system. Before the development of the proposed wireless network management system, we have developed Mobile IPv4 system which consists of HA and FA modules on Linux and MN module on Linux and Windows [10]. We extended the Mobile IPv4 modules to support the Mobile IP MIB completely and some advanced functions such as the tunneled information. Each MN has IEEE 802.11b interface and associated with ORiNOCO AP 2000 product. We used AdventNet SNMP library [11] for SNMP functions and JRobin RRDtool [12] for web-based GUI.

The prototype system gathers information from the common entity information to the entity specific information based on the Mobile IP MIB. However, there are some limitations to just gather the Mobile IP MIB for efficiently managing the node mobility. For example, a mobility management system needs the information on where/when/how much service the mobile node receives during handovers with Mobile IPv4 because this information can be used to track the node mobility. However, network management system has the deficiency for evaluating the amount of serviced data packets through entities of Mobile IPv4, since there is no defined MIB to retrieve that information and already defined received data information in MIB-II includes all received data packets regardless of Mobile IPv4. Thus, we added the tunneled Mobile IPv4 information in the existing interface information MIB (e.g., MIB-II interface group) to classify the amount of serviced traffic from entities of Mobile IPv4. Then we developed the mobile node's mobility tracking function based on tunneled traffic information of mobile nodes and binding cache information of mobility agents.

4.1   Functionalities of the Prototype System

When we implemented the prototype system, we focused on functions related to the node mobility in the AWNMS and the Mobile IPv4. Thus, the prototype system supports the following functionalities; the network topology auto-discovery, binding cache and visitor list management, mobility tracking, and tunneled packet load information function.

- **Network Topology Auto-Discovery Function:** The SNMP-based network topology discovery scheme has been implemented in the "Argus" project [13]. When the network management system needs a topology information, it first gets and analyzes the MIB information of the default router. Based on the MIB analysis, the manager can know node and router information in the network connected to the default router. This operation continues to newly found routers and finally the manager can know the subnet and node information to which subnet it belongs.
- **Binding Cache and Visitor List Management Function:** In the Mobile IPv4, the HA and FA manage information about mobile nodes called the binding cache and visitor list. The binding cache and visitor list management is important to support a node mobility and traffic tunneling in the Mobile IPv4. The prototype system provides management function to access, update, and delete information related to binding cache and visitor list through the SNMP GET/SET methods.
- **Mobility Tracking Function:** The prototype system records the history of the node mobility and periodically retrieves the current location information of the mobile node using the Mobile IP MIB. This information is stored into the Integrated MIB and the active SNMP agent creates a mobility history and pattern. Using the mobility pattern of each node, the SNMP agent can provide the recovery of a connection and the fast negotiation of QoS, security and node profiles for reducing the handover latency.
- **Tunneled Packet Load Information Function:** The tunneling operation in the HA (packet encapsulation) and the FA (packet decapsulation) of the Mobile IPv4 results in processing overhead and affects the system performance. Thus, the tunneled packet load information should be carefully investigated to prevent a malfunction caused by the overloads of Mobility Agents. In the prototype system, the HA and FA gather the tunneled packet size per each mobile node and the network management system shows results to the user through the line plotting chart of transmission rate (bps) as a function of time.

4.2   Architecture of the Manager

We have implemented the active SNMP agent based on the conceptual agent architecture shown in Fig. 3. In this section, we describe the details of manager. The manager consists of five modules as illustrated in Fig. 4; Scheduler, Alive-Checker, SNMP-Requester, Event-Handler, and Mobility-Handler. The Alive-Checker module periodically checks the availability of managed systems using a PING request-response exchange. The SNMP-Requester module periodically sends a SNMP request message to managed systems for gathering the defined MIB information. The interval of these two modules is scheduled by the Scheduler module, which creates a thread on each interval per managed system. The Scheduler module has a thread pool and limits the number of threads executed in parallel. The Event-Handler module is executed when it receives a response of the PING request sent by the Alive-Checker module or the SNMP request message sent by the SNMP-Requester module, and updates the status of the managed system or the corresponding MIB information. The Event-Handler module also transfers the updated information to the Mobility-Handler module. The Mobility-Handler module manages the status and the handover event of mobile nodes. Therefore, the Mobility-Handler module checks whether the information is transmitted from the mobile node or not. If the information is about the mobile node, it records the mobility related information such as a location history of the mobile node. The Mobility-Handler also periodically checks whether the mobile node returns to its home or not.

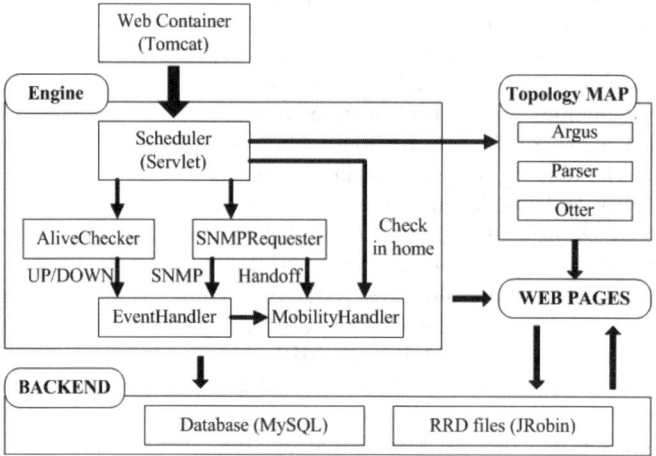

Fig. 4. System architecture of the implemented system

4.3 Implementation Results

The main window of the web-based manager is shown in Fig. 5. In the main window, administrators can view the network map, the list of managed systems and the list of important events. When the detailed information of the managed system is needed, administrators can view the information, as shown in Fig. 6, through selecting a specific managed system in the topology map. Fig. 6 shows a node view page of the FA information. In the node view page, administrators can view the basic information of the FA, visitor list information managed by the FA, AP list information in the FA network, and performance charts representing the number of agent advertisement messages which it sent, agent solicitation messages which it received, and agent advertisement messages which it sent responding to the solicitation message from the mobile node.

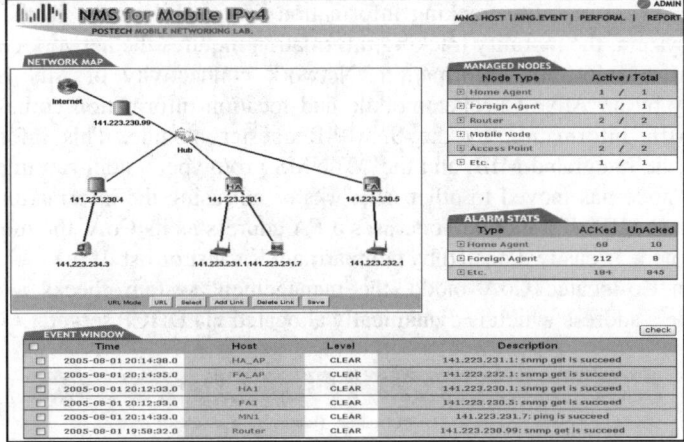

Fig. 5. The main window of the web-based manager

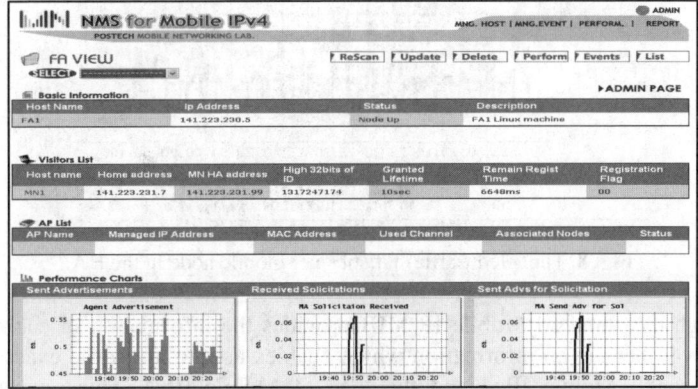

Fig. 6. Node view

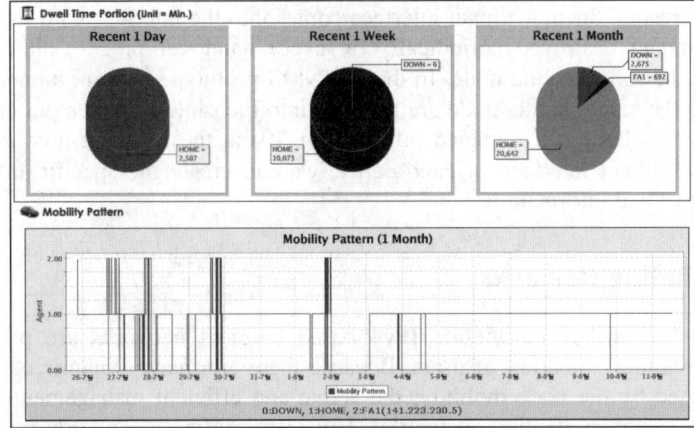

Fig. 7. Mobility tracking view

Fig. 7 shows a mobility tracking information of a mobile node. In the AWNMS prototype system, the mobility tracking information requires the network connectivity information and location information. Network connectivity of MN is checked periodically by an Alive-Checker module and location information comes from the gathered MIB information by the SNMP-Requester module. This information is stored into the Integrated MIB, and the AWNMS prototype system can infer whether the mobile node has moved to other networks or not using the information stored in the Integrated MIB. If a mobile node uses a FA address as its CoA, the movement of the mobile node is easily detected by comparing a FA visitor list. If the mobile node is operated in Co-located CoA mode, the management system checks a change of mobile node's address which is dynamically allocated via DHCP server.

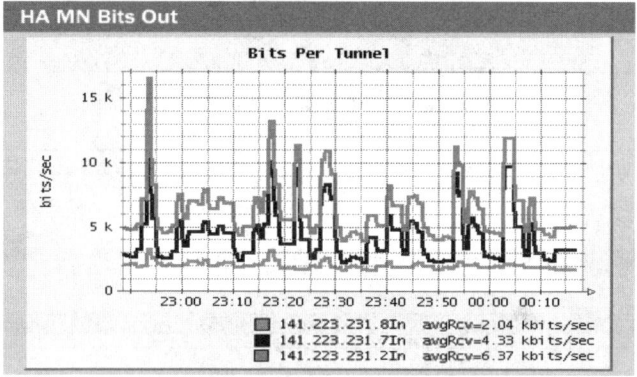

Fig. 8. Tunneled traffic statistics per mobile node in the HA

Although the standard Mobile IP MIB provides the information of each entity of Mobile IPv4, there is no information which enables user to know the exact servicing packets through Mobile IPv4 and the load of Mobile IPv4. Thus, we added the tunneling packet load information to know the exact servicing packets from each agent and the load measurement with Mobile IP MIB. When a tunnel between a HA and FA is created, the new tunnel interface comes into the interface card list. Tunnels are created and destroyed periodically on every handover process although their targets are the same mobile node. In the AWNMS prototype system, tunneled traffic destined to the same mobile node are grouped into the same management entity. Fig. 8 shows the gathering of tunneled information. Using the tunneled forwarding data packets and tunnel interface of each entity, we can create the specific information related to the load information.

5   Concluding Remarks

In wireless networks, the Mobile IPv4 based wireless networks are proposed to support a seamless mobility of users. Because there are many dynamic changes and events caused by the node mobility, the active and efficient management system is required to manage wireless networks. However, ordinary network management systems operate in the passive mode. To overcome this passive characteristic, we have proposed the Autonomic Wireless Network Management System for the management of Mobile IPv4 based wireless networks. The AWNMS has five components (admission control, automatic topology tracking, access selection, mobility tracking, and Integrated MIB) and provides useful functionalities focusing on the node mobility and the communication quality. The autonomic management concept makes the active SNMP agent adaptively manage its managed systems. In the AWNMS, the active SNMP agent enables the autonomic management concept through two components; the self-management and the manager function which is the ability to exchange information with neighbor agents.

We have also designed and implemented the AWNMS prototype system focusing on the events caused by the node mobility. The prototype system has implemented the Integrated MIB using the Mobile IP MIB and MIB-II, and supports a network topology auto-discovery, binding cache and visitor list management, mobility tracking, and tunneled packet load information functions. We have tested the prototype system in a test-bed network and obtained desirable results. Although the prototype system provides limited functionalities of the AWNMS, our design and implementation can be the basis of developing a full AWNMS. In future, we plan to implement other functionalities of the AWNMS based on the prototype system, and prove the feasibility of the AWNMS in the Mobile IPv4 based wireless networks.

References

1. IEEE 802.11, Wireless LAN Medium Access Control (MAC) and Physical Layer (PHY) Specifications, Standard, IEEE, Aug. 1999.
2. IEEE 802.16-2004, Air Interface for Fixed Broadband Wireless Access Systems, Standard, IEEE, Oct. 2004.
3. 3GPP Working Group works in progress (Rel6, Rel7).
4. C. Perkins, "IP Mobility Support for IPv4," RFC 3344, IETF, Aug. 2002.
5. D. Johnson et. al., "Mobility Support in IPv6," RFC 3775, IETF, Jun. 2003.
6. J. Case, M. Fedor, M. Schoffstall, and J. Davin (Eds.), "A Simple Network Management Protocol (SNMP)," RFC 1157, IETF, May 1990.
7. William Stallings, "SNMP, SNMPv2, SNMPv3, and RMON1 and 2," 3rd Ed. Addison Wesley, 1999.
8. D. Cong and M. Hamlen, "The Definitions of Managed Objects for IP Mobility Support using SMIv2," RFC 2006, IETF, Oct. 1996.
9. M. Liebsch et. al., "Candidate Access Router Discovery (CARD)," RFC 4066, IETF, Jul. 2005.
10. Dong-Hee Kwon et. al., "Access Router Information Protocol with FMIPv6 for Efficient Handover and Their Implementation," IEEE Globecom 2005, vol. 6, pp. 3814-3819, 2005.
11. AdventNet SNMP Library, http://snmp.adventnet.com/
12. JRobin, http://www.jrobin.org.
13. Project Argus – Network topology discovery, monitoring, history, and visualization, http://www.cs.cornell.edu/boom/1999sp/projects/Network Topology/topology.html
14. IEEE 802.11F, IEEE Trial-Use Recommended Practice for Multi-Vendor Access Point Interoperability via an Inter-Access Point Protocol Across Distribution Systems Supporting IEEE 802.11 Operation, Standard, IEEE, Jul. 2003.
15. M. Mitchell Waldrop, "Autonomic computing: The technology of self-management," IBM white paper, http://www-03.ibm.com/autonomic/library.shtml, 2003.

A Comparison of Mobile Agent and SNMP Message Passing for Network Security Management Using Event Cases

Ching-hang Fong, Gerard Parr, and Philip Morrow

School of Computing and Information Engineering, Faculty of Engineering, University of Ulster, Coleraine, Co. Londonderry, BT52 1SA, United Kingdom
{fong-c, gp.parr, pj.morrow}@ulster.ac.uk

Abstract. Research has proposed that next generation Mobile Agent (MA) technology will achieve the overall notion of "Zero Touch" network management. The advantages offered by using MA-based Network Management (MANM) include reduction in network traffic, intelligence, automation, fault-tolerance, and robustness. From our previous research, we have developed a security-focused MA framework (MASAM) for achieving active network management. This paper aims to examine the implication of network traffic load when implementing network security management by using MANM through four simulated security attack event cases. Evaluation results indicate that the MANM approach can enhance performance and security when dealing with various security attacks.

Keywords: Network Management, Mobile Agent, Security Architecture, Security Attack Event Case.

1 Introduction

Network management concepts emerged from the need to address problems associated with the increasing complexity and range of networking devices and topologies in the 1980s. Two standards were developed: the Simple Network Management Protocol (SNMP) [1] proposed by the Internet Engineering Task Force (IETF) and the Common Management Information Protocol (CMIP) [2] proposed by the International Standards Organisation (ISO). The dominant SNMP standard and its centralised philosophy are relatively old and pose several issues, especially in regard to performance and security. It requires significant processing capabilities at the central management host and creates unnecessary bandwidth consumption on the network. Thus, it is not sufficient for managing today's narrow bandwidth, mobile or ad hoc formed networks. More importantly, the centralised framework has inherited limitations that leave it difficult to implement security management functions required by the ISO Fault, Configuration, Accounting, Performance, and Security (FCAPS) [3] management framework in a pro-active manner.

A number of research projects [4-6] have proposed that Mobile Agent (MA) technology is the optimal approach, promising "Zero Touch" network management.

G. Parr, D. Malone, and M. Ó Foghlú (Eds.): IPOM 2006, LNCS 4268, pp. 156–167, 2006.

Therefore, it is lightweight, flexible, intelligent, fault-tolerant, and may not require any human intervention. Furthermore, the MA-based management (MANM) concept can more easily implement sets of active and dynamic network management functions. Before MA technology can be adapted for widespread use in network management, however, security issues must be tackled. Our previous research [15] has developed a security focused MA framework for achieving active network management functions.

The rest of this paper is as follows. Section 2 provides essential background knowledge on our previous research and reviews our proposed MA security framework. In section 3, we present the results of our initial experiments on network traffic load generated by SNMP message forwarding and MA migration. Four simulated security attack event cases are defined in section 4, and then used for an evaluation of the MA security framework. Finally, we provide some thought on further work and on concluding results.

2   Background

2.1   Network Management: From SNMP to MANM

The well-known ISO FCAPS management framework defines five Management Functional Areas (MFAs), which are the primary means of supervising the functionalities of network management. With this philosophical framework, SNMP utilises the client-server computation paradigm to achieve the FCAPS MFAs but, of course, neither completely adopted nor in a sophisticated manner. Computer systems have evolved from standalone computation models into a client-sever paradigm. Throughout this evolution, a limited form of code mobility has existed [7]. The catalyst is the concept of "software agent", the superset of MAs. MAs are autonomous computational entities that can suspend and move (including program code, stored data and its state information, such as the execution stack) to another agent-enabled platform on the network, and resume execution, deciding where to go and what to do along the way [8]. The concept of an MA deals with certain special properties that distinguish it from a static program. As commonly agreed by several authors in [9-11], these properties can be classified into mandatory (including Autonomous, Reactive, Proactive, Mobile, Communicative, and Collaborative) and orthogonal (including Learning, Deliberative, and Delegable, but not limited to these alone).

The proposal for the use of MA technology to implement network management is originally based on the idea of reversing the logic with respect to the centralised model. In the other words, management information produced by managed devices (MDs) is periodically transferred to the central management system [12]. Due to the mobility of MAs, management tasks can be partially moved from the manager to the devices, certain micro-management operations locally and migrated back to the manager with the results only. In contrast to the centralised model, MANM can provide optimised performance in low bandwidth, high latency or unreliable network connections. With the notion of Management by Delegation (MbD) [13], MANM offers easier management of today's wireless, mobile or ad hoc formed networks with heterogeneous architectures. The only requirement is to provide a common agent platform.

2.2   A Security Focused MA Framework for Network Management

From our philosophy, a security focused MA framework for achieving secure and active network management (MASAM) consists of four concepts. First, a security requirements model should be defined and kept in mind during the design phase of a MANM system. Moreover, the system should also be fully tested ensuring it conforms to these requirements. Second, a framework of security architectures should be proposed in order to secure each of the management components as the set of baseline security facilities which secure the entire MANM infrastructure. Third, a set of security event response mechanisms (the MA-based security management proposed from the FCAPS framework), should be defined in order to allow recovery of the system from the threats which are impossible to detect or from any unpredictable security compromise. Finally, a certain degree of active defence mechanisms, such as an MA-based Instruction Detection Service (IDS), should be defined. With the notion of intelligence, using MAs to achieve advanced active intrusion detection can facilitate security in both the MANM system and the managed network. Fig. 1 provides an architectural view of the MASAM framework.

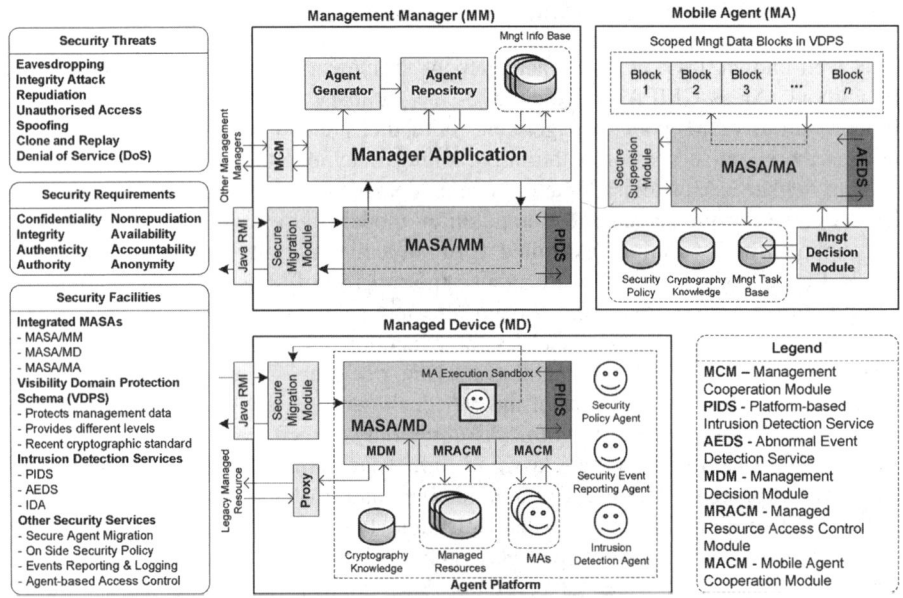

Fig. 1. An Architectural View of the MASAM Framework

In previous papers [14, 15], we presented a security requirement model, CINA, which includes eight security requirements to guide MANM security: there are Confidentiality, Integrity, Authenticity, Authority, Non-repudiation, Availability, Accountability, and Anonymity. With the requirement model identified, we proposed a framework of three integrated MA Security Architectures (MASAs) including one for each of the management managers (MASA/MMs), the managed devices (MASA/MDs), and the management MAs (MASA/MAs). The framework was built from an MA life-cycle model, and with the intention of securing each state of an MA, including execution, suspension, migration, and resumption prior to execution. For reliable high-level security, the framework should be able to integrate with current security standards and techniques. As this framework acts as a baseline protection, the MA should be protected from attacks by malicious agent platforms or other agents; the agent platform should be protected from attacks by malicious agents, and the entire Network Management Infrastructure (NMI) should be protected from attacks by external entities. The proposed MASAs suggest a wide range of security services that include: two sided authentication, resource access control, auditing, encryption and encapsulation, integrity and non-repudiation, sandboxing, secure MA migration, secure data destruction, and intrusion and abnormity detection.

For securing the integrity of a MANM system, there should also be a focus on the protection of management information and functions, and the implications of real-time network management decisions that are based upon such data. While the safe-guard objectives of the proposed MASAs are mainly the managers, the managed devices, and the management MAs, we proposed Visibility Domain Protection Schema (VDPS) [15], which intends to protect the management information being stored or handled physically by the management components of the NMI. This includes are not only the management MAs, but also the managed devices and the management managers. The idea of the VDPS is that the management components are capable of storing their handling management information in the form of a series of Scoped Data Blocks (SDBs). Each SDB will handle a piece of arbitrary management information either encrypted or plaintext and scoped into one class of Visibility Domain (VD), such as the host MA only, specific manager, or entire management domain. The class of VD can be swapped loosely by the SDB holder. The main advantage offered by the VDPS is provision of opportunity that management components are able to communicate to each others directly with encrypted data, which can benefit performance and reduce implementation complexity.

2.3 Classification of Management Mobile Agents

To implement the MANM functions, the MASAM framework introduces two groups of management MAs: Generic Network Management Agent Group (GNMAG) which

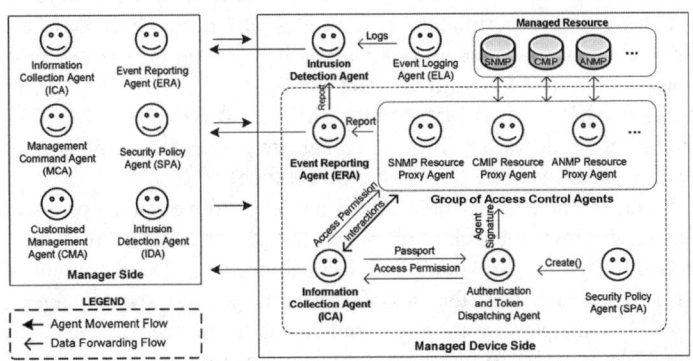

Fig. 2. Interactions for Different Types of MAs Suggested by the MASAM

the MAs use for implementing normal daily network management tasks; and Security Management and Response Agent Group (SMRAG) which the MAs use to achieve security related tasks, such as authentication, or attack response. Additionally, in real-world applications, the MAs from SMRAG should be considered and executed with a higher priority. Fig. 2 illustrates the interactions between each type of MAs, while table 1 discusses the roles of these MA types.

Table 1. A Description of Different Management Mobile Agent Types

MA Types		Roles
Generic Network Management Agent Group		
ICA	Information Collection Agent	Responsible for collecting management information from remote MDs for the manager
MCA	Management Command Agent	Responsible for performing operations on the MDs on behalf of the manager
CMA	Customised Management Agent	Does mixed management tasks of the ICA and MCA
Security Management and Response Agent Group		
SPA	Security Policy Agent	Responsible for managing the security policy (database of rules) of a MD (Generated by and received from the manager)
ATDA	Authentication and Token Dispatching Agent	Responsible for performing MA authentication and access permit dispatching tasks
RACA	Resource-based Access Control Agent	Responsible for providing a common interface for the GNMAG's MAs in order to access current management standards, e.g. SNMP.
SRA	Security Response Agent	Responsible for performing security event response actions in event of security alert is initialised
ERA	Event Reporting Agent	Responsible for staying onto a MD and receiving security-related reports internally
IDA	Intrusion Detection Agent	Responsible for introducing advanced MA-based intrusion detection services for the entire managed network

3   Examining SNMPv3 Messages and Agent Migration

An initial experiment has been conducted in order to estimate, for SNMP, the size of each type of SNMPv3 message sent with AuthPriv (SHA-DES) security parameters set; and, for MANM side, the average weight of network traffic generated by a MA migration. During the experiments, two modes are defined: SNMP and MANM modes. The testbed was formed in our University's research network, and based on two machines: 1) desktop PC with an Intel Pentium 4 3.2GHz CPU, 1GB Ram, and running Windows XP; and 2) laptop with an Intel Centrino 1.5GHz CPU, 768MB Ram, and running Windows XP. Two machines are connected through a 100Mb/s fast Ethernet switch, and using TCP/IPv4 Internet Protocol.

   In order to realise the test for the MANM mode, we developed a prototype MANM system by using the Java Agent DEvelopment (JADE) framework from the Italia Lab. The developed programs are compiled by using Java2 Development Kit (JDK), version 5.0. During each test, the desktop machine acts as the manager, where the simple MANM manager is running and capable of generating a number of management MAs in order to facilitate the tests; meanwhile, the laptop machine acts as the MD, where a simple agent platform is running. For the SNMP mode, two network management products are selected: NuDesign Visual MIBrowser, and NuDesign Multiprotocol (SNMPv3/HTTP) Agent Service. To realise the SNMPv3 message exchanges, a copy of the SNMPv3 agent service is installed on the laptop machine, while the desktop machine sends SNMP requests to and receives SNMP responses from the agent by using the MIBrowser tool. For the result analysis, the packet capturing and analysis tool, Ethereal, is employed. The tool uses a WinPcap driver in order to capture data packets from Windows base environment. Before carrying out each experiment, Ethereal is started on the desktop with a filter that only captures those packets sent to or received from the laptop.

From our experiment results, table 2 shows the average network traffic generated by different kinds of SNMPv3 messages. According to the structure of packets captured by Ethereal, we can conclude that the overhead size of each SNMPv3 message sent through TCP/IPv4 is 122 bytes. According to the results from the MA migration experiments, we can estimate that the network traffic generated by an MA migration is 1120 bytes. Understanding the average size of each type of SNMP message and an MA migration can benefit to achieve evaluation of the traffic load implication on different approaches when dealing with security attack events later on.

Table 2. Estimated Size of SNMPv3 Message Types

Message Type	Traffic (Bytes)	Message Type	Traffic (Bytes)
GetRequest	199	Response(GetRequest)	204
GetBulkRequest	200	Response(GetBulkRequest)	513
SetRequest	221	Response(SetRequest)	220
InformRequest	213	Response(InformRequest)	214
Trapv2	211		

4   Defining Security Attack Event Cases

A security attack scenario (event cases) approach has been chosen in order to evaluate the network traffic load implications on MA and SNMP based network management systems. An event case involves a possible security attack which is performed by one intruder and targeted to one or more components. To verify the possibility for assessment of the selected cases, some selection criteria have been setup: 1) the simulated attack should be detectable by both MANM and SNMP modes; 2) the case should involves three definite stages (attack, detection, and recovery); 3) the case may show scalability implication when testbed environment is growing. In this paper, we have chosen four simulated event cases for our experiments: Case 1. Network footprinting attack; Case 2. Virus infection attack; Case 3. Message/MA replay attack; and Case 4. Man-in-the-middle attack. Fig. 3 shows a simulated network environment for implementing these selected event cases. It should be noted that manager 1 (MM1) manages all elements within the Management Domain A (DmA) except the intruder 1 (E1); manager 2 (MM2) manages all elements within the domain B (DmB) except the Intruder 2 (E2). None of the elements outside of the DmA and DmB will be managed by either MM1 or MM2. The basic NMI for this testbed is SNMPv3 protocol with component support for SNMPv1. For MANM mode, every MD is running an MA platform while the MASAM framework is adopted.

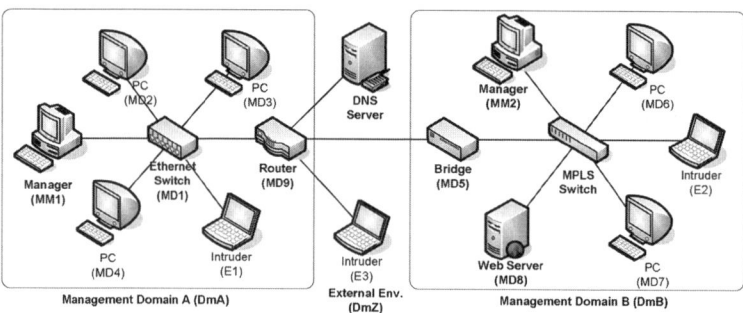

Fig. 3. Simulated Testbed Environment for Implementing the Security Attack Event Cases

4.1 Network Footprinting Attack

The basic goal of a network footprinting attack is to earn as much knowledge about the network and usually the first step in attacking any network which is to figure out what to attack. The management information handled by network management systems is most likely this kind of "useful" information. In our simulated event case, the intruder E1 launches a network footprinting attack on a switch MD1 in order to compromise the managed network DmA later on. The assumption is made that the intruder has already acquired unauthorised "Read" access permission on the MD1 by using other means. In SNMP mode, the intruder sends a number of SNMPv3 requests to the MD1 in order to obtain a full-set of MD1's MIB data (totally 242 objects); In MANM mode, the intruder dispatches a malicious MA to do the same thing. Fig. 4 shows the detailed message passing steps of this event case. Observation shows that the MANM approach can reduce much of network traffic in the attack stage, and generate no network traffic for remain stages because decisions are made locally.

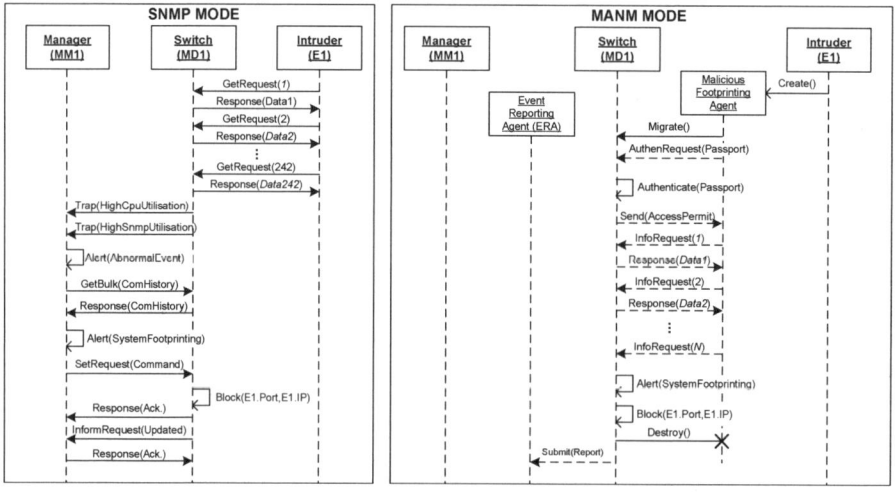

Fig. 4. Sequence Diagram for Illustrating a Network Footprinting Attack Event Case

4.2 Virus Infection Attack

A virus attack can be dangerous to not only the infected victim but also other nodes the victim has communicated with. In our simulated event case (as shown in Fig. 5), the intruder E3 launches a virus infection attack on a machine MD2. The intruder firstly compromises this machine by using other means and infects it with a virus. The virus aims to: 1) infect every accessible file on the machine with a copy of this virus, and 2) broadcasts a joke message (20 bytes) to every host (totally 9 nodes) the victim knew in every 50ms. Due to this, the disk and CPU loads and network activities will become abnormally high. This abnormal symptom provides an indicator for a manager MM1 that the machine may be infected. It is assumed that this symptom has been discovered after 30 seconds of the attack (i.e. after 5400 messages sent). Two steps of response action the manager will take: 1) try to shutdown the infected machine; and 2) disable the machine's port access from the switch MD1. Similar to the footprinting attack, this case shows that the MANM approach generates less traffic in general. However, in the recovery stage, this approach may create higher traffic if several MA migrations are required.

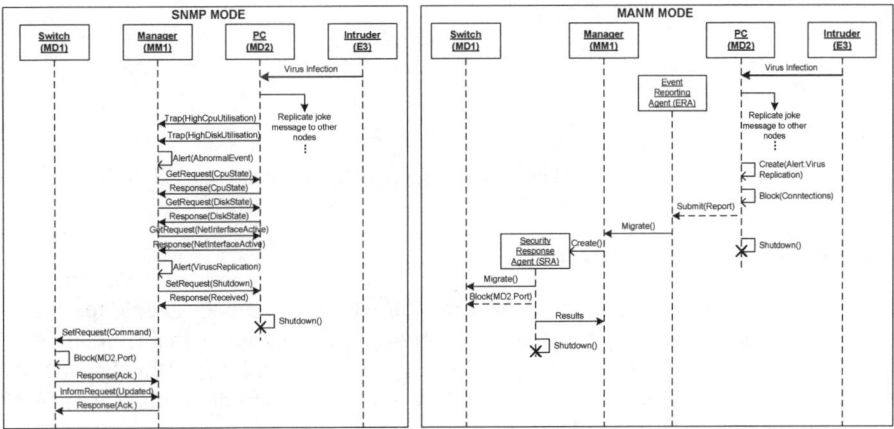

Fig. 5. Sequence Diagram for Illustrating a Virus Infection Attack Event Case

4.3 Message/Agent Replay Attack

In a replay attack, a valid message or data transmission is maliciously or fraudulently repeated by an intruder. In our simulated event case, the intruder E2 launches a Message/Agent attack on a machine MD7. The first step of this attack is to place a sniffer onto the network DmB in order to obtain a transmitting SNMP message or MA. Eventually, a copy of (for SNMP mode) *SetRequest(Restart)* message (for MANM model, a MCA MA) has been captured. After that, the intruder is able to replay. For the SNMP mode, the SNMPv3 is still vulnerable if the sniffed message is being replayed within 160ms. Thus, the victim will be continuously restarted (it is assumed 5 times) until the message is invalid. In contrast, since the MANM mode implements the MASAM security framework which protects systems from MA-based replay attack, the cloned MA will not be able bypass the authentication mechanism. It is assumed that the replayed MA will be failed on authentication stage and destroyed immediately. Fig. 6 shows the detailed message passing steps of this event case. Observation shows that the MANM approach can protect the system from the initial stage of the attack, and generate no network traffic in the detection stage.

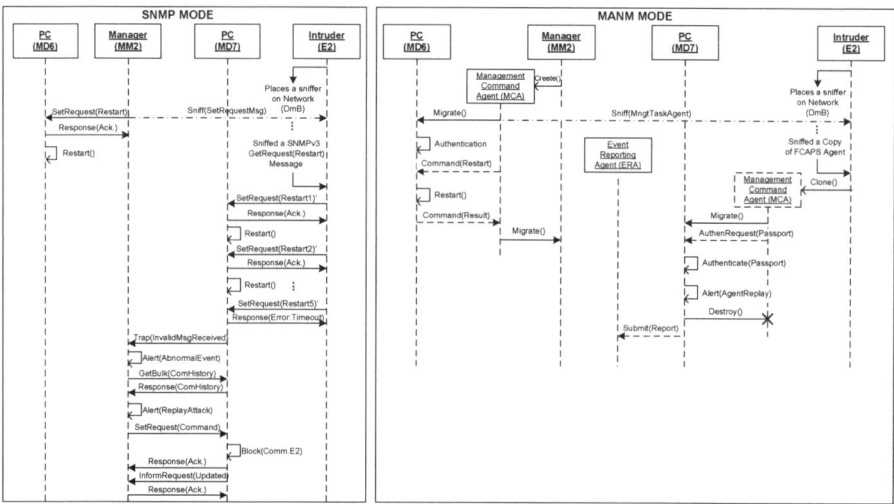

Fig. 6. Sequence Diagram for Illustrating a Message/Agent Replay Attack Event Case

4.4 Man-in-the-Middle Attack

Man-in-the-Middle (MiM) attack is an attack in which an intruder is able to receive and modify messages or data exchanges between two parties without either party knowing. In our simulated event case (see Fig. 7), the intruder E1 launches a MiM attack between two machines, MD3 and MD4. Firstly, the intruder initialises an ARP cache poisoning attack [16] (4 fake ARP responses sent, 60 bytes for each) onto a switch MD1 in order to utilise this switch to redirect all traffic which are designated for the victims to the intruder's machine. When a message has been wrongly redirected, the intruder modifies this message for malicious purposes and then returns back to target recipient. The number of messages exchanging between the victims will be doubled.

This compromise will be detected when, for the SNMP model, the manager MM1 performed a periodical ARP table integrity check on the DmA; while, for the MANM model, an intrusion detection MA (travels around the DmA and) detected some inconsistent ARP entries. It is assumed that 50 messages (50 bytes for each) have been manipulated before the attack is detected. Finally, the manager recovers the entire ARP table (in our case, this was 160 bytes. This will obvious change depends on the number of nodes on the network) of the switch. The intrusion detection MA may create some traffic on migrations. However, this MA provides more advanced and intelligent detection services not only for ARP table but also abnormal events.

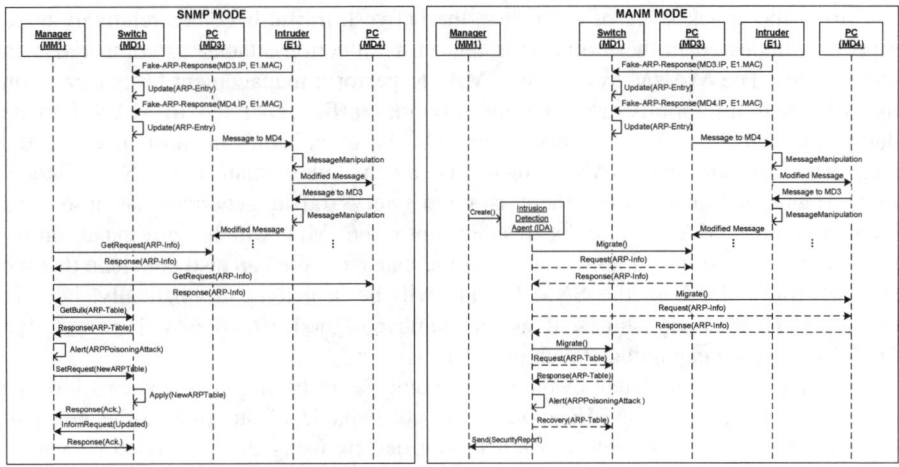

Fig. 7. Sequence Diagram for Illustrating a Man-in-the-Middle (MiM) Attack Event Case

5 Summary and Evaluation

Evaluation of the performance implication for adopting new kind of techniques in network management can be focussed on three factors: 1) additional network traffic, and 2) overhead processes on the manager and 3) the MD. Table 3 summarises the volume of total network traffic generated by each attack event case in different stages.

Table 3. Summarised Network Traffic Generated by Each Security Attack Event Case

Case	Stage	SNMP Mode	Traffic*	MANM Mode	Traffic*
1.	*Attack*	GetReqRep x 242	97526	AgentMove	1120
	Detection	Trap x 2 + GetBulkRep	1135	-	0
	Recovery	SetReqRep + InfoReqRep	868	-	0
2.	*Attack*	JokeMsg x 5400	108000	JokeMsg x 5400	108000
	Detection	Trap x 2 + GetReqRep x 3	1451	-	0
	Recovery	SetReqRep x 2 + InfoReqRep	1309	AgentMove x 2	2240
3.	*Attack*	SetReqRep x 5	2205	AgentMove	1120
	Detection	Trap + GetBulkRep	924	-	0
	Recovery	SetReqRep + InfoReqRep	868	-	0
4.	*Attack*	FakeArp x 4 + (50 x 50)	2740	FakeArp x 4 + 50 x R	2740
	Detection	GetReqRep x 2 + GetBulkRep	1519	AgentMove x 2	2240
	Recovery	SetReqRep + InfoReqRep + 160	1028	AgentMove	1120

Remarks:
GetReqRep = Size of a pair of SNMPv3 GetRequest and Response(GetRequest) Messages
GetBulkRep = Size of a pair of SNMPv3 GetBulkRequest and Response(GetBulkRequest) Messages
SetReqRep = Size of a pair of SNMPv3 SetRequest and Response(SetRequest) Messages
InfoReqRep = Size of a pair of SNMPv3 InformRequest and Response(InformRequest) Messages
AgentMove = Total Network Traffic generated by an MA Migration
JokeMsg   = Size of a Joke Message Sent by the Infected Host
FakeArp   = Size of a Fake ARP Response Message
* The unit for measurement of network traffic is in number of Bytes generated

During the attack stage of each case, the network traffic becomes relatively high, especially appeared in the footprinting and virus infection attacks, as compared with other stages. The MANM system uses MAs to perform management tasks locally on the MD, MA migrations are the only network traffic generated in MANM mode during the detection and recovery stages. On the other hand, in most of cases, the recovery cost for the MANM mode is much lower than the SNMP mode. Furthermore, in some cases, there is no network traffic generated because of a security policy MA is already resided onto the MD which can make on-fly management decisions without contacting the manager. We can also conclude that the network traffic load on the SNMP mode will be scaled up dramatically, both in detection and recovery stages, if as the number of nodes increases. However, the MANM approach can reduce much of traffic.

Through these simulated event cases, some security implications can also be discovered. Firstly, the SNMP approach is not capable of stopping some kinds of attacks until an attack symptom has been notified by using the responded data from periodical sent state requests. For example, in the footprinting attack case, the manager sent periodical communication history requests to the MD and received responses, and then an alert is created on the manager side. However, with the MANM approach, it is because the security policy MA resides on the MD it can alert itself and destroy the potential malicious MA immediately. Thus, no information will be enumerated by the intruder. Secondly, with MANM approach, some security attacks may be prevented in the initial stage. For example, in the replay attack case, the replayed MA can not bypass the authentication and will be destroyed. However, due to the design flaw of the SNMPv3, several replays can be successful until the message itself becomes invalid. Thirdly, with the MANM approach, the security policy can be easily implemented locally on and customised to a MD. The tightness of the policy can be changed dynamically based on current situations. In contrast, the SNMP approach implements a centralised security policy database for every MD on the managed network. This approach is lacks flexibility and requires all decision making to be based on the central host.

6   Conclusions and Further Work

This paper provides an architectural review of our MASAM framework in supporting secure and active MA-based network management. A classification of the MA family for implementing this framework has also been discussed. Four simulated security attack event cases have been defined in order to have a study on the implication of network traffic load within both SNMP and MANM approaches. The event cases have been evaluated through observation and statistical analysis. Positive evaluation results have been observed which shows that the MANM approach (conforming to the MASAM security framework) can enhance performance and security when dealing with security attack events.

The future work on this research is to model these event cases in more detail and at different scales. The event cases should be enlarged from currently one-to-one based scenario into a multicast or broadcast based attack environments. The objective of this modelling work is to seriously evaluate the traffic load implications on the network, the managers and the MDs in large heterogeneous network environment in order to obtain more complete results. The intention is to examine the application of our security models in the converged networks which promotes development of next generation networks in the future.

References

[1] D. Harrington, R. Presuhn, B. Wijnen, "An Architecture for Describing Simple Network Management Protocol (SNMP) Management Frameworks", RFC3411, IETF, 2002

[2] ISO/IEC 9596-1:1998, "Common management information protocol - Part 1: Specification", Open Systems Interconnection (OSI), October 2003

[3] ITU-T X.700, "X.700: Management framework for Open Systems Interconnection (OSI) for CCITT applications", ITU-T Recommendations, September 1992

[4] R. Stephan, P. Ray, and N. Paramesh, "Network management platform based on mobile agents", International Journal of Network Management, Vol. 14, 2004, pp.59-73

[5] Timon C. Du, Eldon Y. Li, and An-Pin Chang, "Mobile Agent in Distributed Network Management", Journal of Communications of the ACM, 2003, Vol. 46, Iss. 7, pp.127-132

[6] S. Papavassiliou, A. Puliafito, O. Tomarchio, and J. Ye, "Mobile agent based approach for efficient network management and resource allocation: framework and applications", IEEE Journal on Selected Areas in Communications, Vol. 20, Iss. 4, 2002, pp.858-872

[7] Hyacinth S. Nwana, "Software Agents: An Overview", Knowledge Engineering Review, Vol.11, Iss.3, October 1996, pp.205-244

[8] OMG, "Agent Technology, Green Paper", OMG Document agent 00-09-01, Version 1.0, Agent Platform Special Interest Group, Object Management Group (OMG), September 2000, Available online at http://www.objs.com/agent/index.html

[9] D. Gavalas, D. Greenwood, M. Ghanbari, and M. O'Mahony, "An infrastructure for distributed and dynamic network management based on mobile agent technology", IEEE International Conference on Communications, 1999, pp.1362-1366

[10] H. Reiser and G. Vogt, "Threat Analysis and Security Architecture of Mobile Agent based Management Systems", Proceedings of NOMS 2000, IEEE/IFIP Network Operations and Management Symposium, Honolulu, Hawaii, USA, April 2000

[11] Kun Yang, Alex Galis, Telma Mota, and Angelos Michalas, "Mobile Agent Security Facility for Safe Configuration of IP Networks", MANTRIP, the EU IST project, 2002

[12] M. Baldi, S. Gai, and G. Picco, "Exploiting Code Mobility in Decentralized and Flexible Network Management", Proceedings of the First International Workshop on Mobile Agents (MA97), Berlin, Germany, April 1997

[13] Symeon Papavassiliou, Antonio Puliafito, Orazio Tomarchio, and Jian Ye, "Mobile Agent-Based Approach for Efficient Network Management and Resource Allocation: Framework and Applications", IEEE Transactions on Systems, Vol.20, Iss.4, May 2002

[14] Ching-hang Fong, Gerard Parr, and Philip Morrow, "Security Implications of Mobile Agent based Network Management: a Review", EPSRC PostGraduate Symposium on the Convergence of Telecommunications, Networking and Broadcasting, 2005, pp.326-332, Available online at http://kenbane.infc.ulst.ac.uk/~ching/paper/PGNet2005-SIMANM.pdf

[15] Ching-hang Fong, Gerard Parr, and Philip Morrow, "MASAM: A Mobile Agent Based Security Framework in Supporting Active Communications Network Management", Technical Paper, Faculty of Engineering, University of Ulster, UK, April 2006, Available online at http://kenbane.infc.ulst.ac.uk/~ching/paper/TechPaper2006-MASAM.pdf

[16] D. Bruschi, A. Ornaghi, and E. Rosti, "S-ARP: a secure address resolution protocol", Proceedings of 19th Annual Computer Security Applications Conference, 2003, pp.66-74

Principles of Secure Network Configuration: Towards a Formal Basis for Self-configuration

Simon N. Foley[1], William Fitzgerald[2], Stefano Bistarelli[4,5],
Barry O'Sullivan[1,3], and Mícheál Ó Foghlú[2]

[1] Department of Computer Science, University College Cork, Ireland
s.foley@cs.ucc.ie
[2] Waterford Institute of Technology, Ireland
(wfitzgerald, mofoghlu)@tssg.org
[3] Cork Constraint Computation Centre, University College Cork, Ireland
b.osullivan@cs.ucc.ie
[4] Dipartimento di Scienze, Università "G. D'Annunzio" di Chieti-Pescara, Italy
[5] Istituto di Informatica e Telematica, CNR, Pisa, Italy
bista@sci.unich.it

Abstract. The challenge for autonomic network management is the provision of future network management systems that have the characteristics of self-management, self-configuration, self-protection and self-healing, in accordance with the high level objectives of the enterprise or human end-user. This paper proposes an abstract model for network configuration that is intended to help understand fundamental underlying issues in self-configuration. We describe the cascade problem in self-configuring networks: when individual network components that are securely configured are connected together (in an apparently secure manner), a configuration cascade can occur resulting in a mis-configured network. This has implications for the design of self-configuring systems and we discuss how a soft constraint-based framework can provide a solution.

1 Introduction

Today's management of the telecommunications and enterprise network infrastructure has become fundamentally more complex than it has been in the past. Network elements have expanded to include functionality to cater for a greater variety of applications and across more layers within the network infrastructure [1]. In order to manage the current complexity of networks, both telecoms and enterprise providers use the support of network management systems (NMS) that provide operational and maintenance capabilities at various levels throughout the network [2]. The process of efficiently deploying a complex system of network services on a complex system of network system nodes and having to manage it thereafter is tedious and error prone. These systems are now becoming overwhelmed by the demands placed on the network by its customers [3].

Current research is investigating new avenues to automate the process of NMS to autonomously handle the labour intensive and error prone configurations.

G. Parr, D. Malone, and M. Ó Foghlú (Eds.): IPOM 2006, LNCS 4268, pp. 168–180, 2006.

Autonomic computing is one such area that attempts to address the challenges posed on current network management systems. The term *Autonomic Computing* was coined by IBM in 2001 and it is motivated by how the human nervous system can react to its environment. The autonomic human nervous system frees the conscious mind from the burden of having to deal with vital but lower-level functions. Similarly, autonomic computing is intended to free system administrators from complex management and operational tasks [4,5]. This area revolves around self-management, self-configuration, self-protection and self-healing in accordance with the high level objectives of the enterprise or human end-user [6]. The characteristic of self-(re)configuration plays a large part in ensuring that network devices and its hosted services are configured correctly and, more importantly, in a secure manner.

The contribution of this paper is an abstract model that defines objectives that a self-configuring network entity should uphold. A valid configuration is defined in terms of the quality of the configuration of the underlying components. Using this model we demonstrate a configuration cascade problem, whereby the quality of the security of overall configuration is not just based on the quality of the individual components, but also on how they interoperate. This has practical implications for the design of self-configuring systems. In this paper we are primarily concerned with exploring underlying principles for self-configuration; how these issues relate to the implementation of network management architectures [7,8,9,10,11,12] is beyond the scope of this paper.

The paper is organized as follows. Section 2 outlines an architecture for autonomic network nodes. In Section 3, a formal model is proposed that is intended to characterise the configuration objectives for these nodes. Section 4 describes a metric that is used to specify how effective a node configuration is at achieving its objectives. Using this metric, Section 5 demonstrates that achieving a valid configuration is more difficult than simply ensuring the validity of the configuration of individual components and their immediate connections.

2   Network Node Configuration

Figure 1 depicts the basic structure of a node. A node has an *ability* to perform certain tasks. For example, a router might have the ability to host multiple types of router specific services/protocols to govern its network segment depending on its role within the network at a given moment in time. Nodes assign *contracts* to the services that it hosts. Each node is capable of hosting different types of services but under the self-managed constraints of the node.

Contracts are advertised to the network about a node's ability to offer a particular type of service hosting. A node can migrate or clone, for redundancy reasons, a service to the advertising node based on the contract requirement meeting the needs of the service. Similarly, a node can also advertise to the network the need to off-load certain services and can do so by advertising a service contract in which each node available to host services can decide if it can provide for the service requirements based on its individual policy.

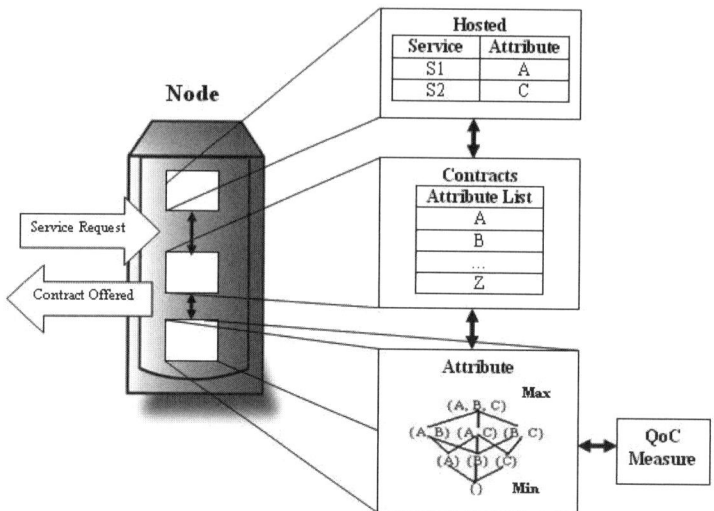

Fig. 1. Architecture of a Network Node

3   A Model of Configuration

A network configuration is defined in terms of a graph of interconnected nodes that host various application-level services. Let *Node* define the set of all possible systems, routers, and so forth that comprise a network. Each node is configured to provide a variety of resources to the services that it hosts. In addition to storage resources such as access to file-systems and databases, hosts may also be configured to offer the use of system-level services/applications, for example, IPv6, SOCKs authentication, SNMP, FTP, and so forth. These are characterised as the *attributes* of a node. Informally, attributes represent anything that a node can offer and/or do.

Let *Attribute* represent the set of all possible attributes of nodes. We use s-expressions [13] to define attributes. This is a Lisp-like notation for naming entity characteristics. Examples of attributes described as s-expressions include:

```
(file "/tmp/foo.bar" ("read" "write"))
(port (1024 1025) ("connect" "listen"))
(app "SSH") (app "Kerberos") (app "POP") (proto "IPv6")
(diskquota "diskA" "100M")
(cpuquota "slice1" "10-MFLOPS") (cpuquota "slice2" "15-MFLOPS")
```

At its most basic level of interpretation, an attribute is not unlike a capability or permission: holding the permission permits a service to access the object in the prescribed way. This interpretation is extended to include anything that a node might make available to services that it hosts, e.g. a disk quota of 100MBytes, or a real-time system that offers a service a 10MFLOPS time-slice.

We assume that the set of attributes forms a partial order under $\leq$. Given $a, b : Attributes$, then $a \leq b$ is interpreted to mean that the attribute b implies

a, that is, if a node offers b, then it implicitly offers a, and similarly, if a service holds attribute b then it also holds attribute a. For example, we would expect the following to hold:

$$(\texttt{file"/tmp/foo.bar"("read")}) \leq (\texttt{file"/tmp/foo.bar"("read""write")})$$

We further assume that this partial order defines a lattice ($Attributes, \leq$, $\sqcup, \sqcap, \perp, \top$), with lowest upper bound operator $\sqcup$, greatest lower bound operator $\sqcap$ and unique lowest and unique highest attributes $\perp$ and $\top$, respectively. In principle, this is not an unreasonable assumption, as in the worst-case, a power-set lattice (with ordering defined by subset and lowest upper bound defined by union) of s-expressions can be used to represent the set of attributes. In this way, the lowest upper bound provides a meaning for the combinations of attributes. Figure 2 gives an example of a powerset lattice of attributes.

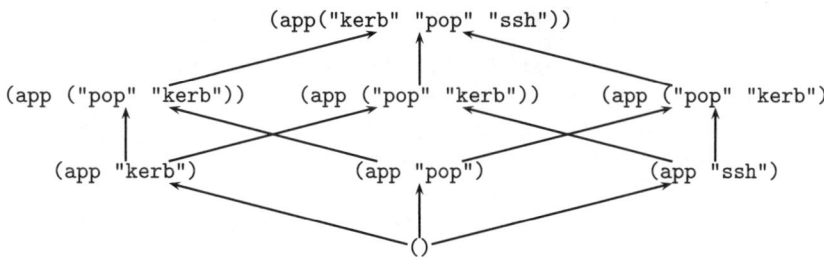

Fig. 2. Sample Powerset Lattice of Attributes

Services. Let *Service* represent the set of all possible services that are to be hosted on network nodes. A dependency relation is defined for services

$$depends : Service \leftrightarrow Service$$

whereby $depends(s_1, s_2)$ means that service s_1 uses service s_2.

In order to operate properly, a service requires a minimum collection of attributes to be available on its hosting node. We define

$$minSvcAttrib : Service \rightarrow Attribute$$

whereby, given service s, then $minSvcAttrib(s)$ defines the minimum attribute required by s. For the sake of simple exposition, we assume that if a service requires attributes $a, b, c, \ldots$, then its minimum requirement is defined by the join of these attributes, that is, $(a \sqcup b \sqcup c \sqcup \ldots)$. For example, based on the attribute lattice in Figure 2, a service s that provides ssh'ified email has requirements $minSvcAttrib(s) = (\texttt{app("pop" "ssh")})$.

Nodes. Nodes are configured to host services. This is specified as:

$$hosts : Node \rightarrow (Service \rightarrow Attribute)$$

whereby, the services that a node n hosts is specified by $hosts(n)$ and this defines a function that maps these services to their assigned attributes. For example, $hosts(n)(s)$ defines the attribute that is provided by the node n to the service s that it hosts. We require that the attributes offered by a node are sufficient for the service, that is, for all nodes n and services $s \in \mathrm{dom}(hosts(n))$,

$$minSvcAttrib(s) \leq hosts(n)(s)$$

The function from *Service* to *Attribute* is a partial injection: the domain of $hosts(n)$ defines the services hosted and no two hosted services share the same attribute. If two different services require similar node attributes, then we assume that the corresponding attribute s-expressions contain sufficient information to distinguish the attribute as offered to different services. For example, node n allows service s to listen on port 1024 over IPv6 using a 10/100 Ethernet card.

```
hosts(n)(s)=(svc s (port 1024 "listen")
             (proto "IPv6") (hw "eth 10/100"))
```

In this way, these attributes may be thought of as node *contracts* that have been assigned by the node to the services that it hosts.

A node is limited by the resources, system-level applications, and so forth, that it can potentially offer to any services that it hosts. We define

$$maxAttrib : Node \rightarrow Attribute$$

whereby, $maxAttrib(n)$ is the maximum attribute that node n can offer. This sets an upper bound on the contract/attributes defined by $hosts$. For any node n we require that the combined contracts assigned as $hosts(n)$ do not exceed $maxAttrib(n)$, that is,

$$\sqcup\{s : Service | s \in \mathrm{dom}(hosts(n)) \bullet hosts(n)(s)\} \leq maxAttrib(n)$$

Note that we use a Z-like notation (such as the set specification in comprehension above) to specify requirements. Recall that the function defined by $hosts(n)$ is an injection and, therefore, the join ($\sqcup$) of the set of attributes assigned, can be used to effectively represent a summation of the contracts issued.

For symmetry, we also define the minimum attributes hosted by a node.

$$minAttrib : Node \rightarrow Attribute$$

whereby, $minAttrib(n)$ is the minimum attribute that node n is willing to offer. For example, a multiprocessor cluster might only offer contacts to services with high computational or resource requirements; a node with an SSL hardware accelerator may only be willing to host services that need to use SSL. Thus, we require the following to hold.

$$minAttrib(n) \leq \sqcap\{s : Service | s \in \mathrm{dom}(hosts(n)) \bullet hosts(n)(s)\}$$

If a node is not concerned with enforcing a minimum contract then we define $minAttrib(n) = \bot$.

Table 1. Sample Node attribute bindings

n :	Alice	Bob	Clare
$minAttrib(n)$	()	(app "pop")	(app "kerb")
$maxAttrib(n)$	(app "pop")	(app ("pop" "kerb"))	(app ("kerb" "ssh"))

Example 1. Table 1 provides sample attribute bindings for nodes based on the attribute lattice defined in Figure 2. Node `Alice` runs sendmail and is configured to offer, at most, only POP to its hosted services. The minimum attribute offered by `Alice` is () ($\bot$) and, therefore, she is also willing to host services that do not need access to any of these resources. Node `Bob` is intended as a dedicated mail node and, therefore, is only willing to host services that intend to use POP (with or without the use of Kerberos). Finally, `Clare` is a host that supports kerberized services that may also require strong authentication services (SSH) for incoming requests that are non-kerberized. $\triangle$

A network configuration is a collection of interoperating nodes that host various services. For the purposes of this paper we consider interoperation simply in terms of links between nodes. We define

$$link : (Node \times Attribute) \leftrightarrow (Node \times Attribute)$$

whereby, $link((n1, a1), (n2, a2))$ means that there is a bi-directional link connecting node $n1$ and node $n2$, that allows a service (on $n1$) assigned attribute $a1$ to interoperate with a service (on $n2$) that is assigned attribute $a2$. We assume that $minAttrib(n1) \leq a1 \leq maxAttrib(n1)$ and $minAttrib(n2) \leq a2 \leq maxAttrib(n2)$. For example, we could have a link from (`Clare`, (app kerb)) to (`Bob`, (app (pop kerb))), reflecting support for a Kerberos-based single sign-on for services between `Clare` and `Bob`.

We define, $link^+$ as the transitive closure of $link$ and it represents all the (assumed traversable) possible paths in the network configuration. Whatever the network configuration may be, we require that services that depend on each other are accessible to each other, that is, for all nodes $n1$ and $n2$, and services $s1 \in \text{dom}(hosts(n1))$ the and $s2 \in \text{dom}(hosts(n2))$ then the following holds.

$$depends(s1, s2) \Rightarrow link^+([n1, hosts(n1)(s1)], [n2, hosts(n2)(s2)])$$

This is a simple interpretation of interoperation that assumes that the existence of a path between nodes implies reachability; it is adequate for our current purposes. Future research will extend the model to consider how the attributes available to hosts may influence reachability along these paths.

4   Quality of Configuration

Different systems can achieve varying degrees of 'quality' in their configurations. For example, a Security Enhanced (SE) Linux system that is configured to support different service proxies in separate protection domains could be considered to be a 'better quality' configuration than a standard Windows operating system-based workstation hosting similar services.

Quality. Let *Quality* define a set of quality measures that are partially ordered according to ≤. For example, security evaluation criteria use *assurance levels* to provide comparisons of 'quality' between different systems; the Common Criteria [14] uses evaluation levels $E1 < E2 < \cdots < E6$ to compare different levels of assurance. In this case, the intention is that one can have more confidence in a system with a higher assurance level to enforce its security requirement.

We argue that the *Quality* measure can be used to represent other measures of interest. For example, $lo < med < hi$ might represent levels of quality of service that are achievable by a node. Quality can be regarded as addressing the risk of failure: a high risk of failure requires a high quality configuration, while a low quality configuration is sufficient if there is low risk of failure. We define,

$$rating : Node \rightarrow Quality$$

where, $rating(n)$ gives the quality rating of a node n. In this paper we do not prescribe *how* a quality evaluation of a node might be done, rather, once a measure is available then we are interested in ensuring that its use is consistent and traceable within a configuration.

Quality of Configuration Policy. A quality of configuration policy is defined in terms of the attributes that nodes offer. A firewall node that provides protection domains for mail and news proxies should be of high quality: we want to be sure that a failure of one proxy cannot interfere with the other on the system. A switch that routes ADSL and/or Gigabit network traffic should have high quality in its design and configuration: we want to be sure that it will not flood the ADSL side, while ensuring proper speeds for Gigabit-only traffic. In these cases, quality depends on the attributes that the node can offer to the services that it hosts. We define the quality of configuration policy in terms of

$$minQual : Attribute \times Attribute \rightarrow Quality$$

where, $minQual(a, b)$ is the minimum acceptable quality of configuration for a node that offers contracts that range from minimum a to maximum b.

Example 2. Consider the attributes defined by the lattice in Figure 2, and security *Quality* ordering $lo < med < hi$. Suppose that a node is to host services that require attributes ranging from a minimum of $\bot$ to a maximum of (`app` `"pop"` `"kerb"` `"ssh"`). POP services can be relatively easily compromised via password sniffing. Therefore, we need confidence that if the POP service is compromised then it will be difficult for the attacker to, in turn, compromise the system and obtain, for example, copies of the locally stored Kerberos session keys. In practice this could be achieved using a system that provides protection domains that can sandbox/constrain a service to just the attributes that it requires. In this case, we define

$$minQual((), (\texttt{app ("pop" "kerb" "ssh")})) = hi$$

This means that a system that is configured to offer this range of attributes must have a high quality configuration.

We conjecture that nodes providing high quality configuration will be expensive, and therefore, if possible, it is preferable to configure networks from multiple cheaper and lower quality components. One TCB-subset-style strategy is to use nodes that offer only limited capability and may, therefore, be of lower quality. For example, there is lower risk associated with a node thats offers only a POP capability; similarly, there is lower risk associated with a node that offers only SSH. In general, there is a lower risk when nodes/systems are deployed to offer a smaller number of service attributes than when large combinations of attributes are possible. Therefore, we define, $minQual((), (\texttt{app "kerb"})) = lo$, and so forth.

A node that hosts services requiring only pairs of these attributes is defined to require medium quality configuration; for example,

$$minQual((\texttt{app "pop"}), (\texttt{app ("pop" "kerb")})) = med$$
$$minQual((\texttt{app "kerb"}), (\texttt{app ("kerb" "ssh")})) = med$$

It follows, that if $rating(\texttt{Alice}) - low$ and $rating(\texttt{Bob}) - rating(\texttt{Clare}) = med$, then the nodes meet the quality of configuration requirement. $\triangle$

Example 3. Network nodes are configured with support for ADSL, 10/100 Ethernet and 100/1000 Gigabit Ethernet. We define a simple ordering over these attributes as follows:

$$(\texttt{hw "ADSL"}) < (\texttt{hw "eth 10/100"}) < (\texttt{hw "eth 100/1000})$$

The node/switch $\texttt{Eric}$ is configured to route ADSL and 10/100 traffic and, therefore, the attributes that it offers to a service are constrained as follows.

$$minAttrib(\texttt{Eric}) = (\texttt{hw "ADSL"})$$
$$maxAttrib(\texttt{Eric}) = (\texttt{hw ("ADSL" "eth 10/100")})$$

The node $\texttt{George}$ is configured to route only 10/100 and 100/1000 traffic:

$$minAttrib(\texttt{George}) = (\texttt{hw "eth 10/100"})$$
$$maxAttrib(\texttt{George}) = (\texttt{hw ("eth 10/100" "eth 100/1000")})$$

A switch configured to route both ADSL and 10/100 traffic A quality switch attempts to avoid flooding the ADSL connections and slowing the 10/100 connections, while a low-quality switch cannot ensure this. Using the same quality ordering as the previous example, we define

$$minQual((\texttt{hw "ADSL"}), (\texttt{hw ("ADSL" "eth 10/100")})) = med$$
$$minQual((\texttt{hw "eth 10/100"}), (\texttt{hw ("10/100" "eth 100/1000")})) = med$$

If a switch routes only single-speed traffic, then it need not have high quality, as flooding, etc., is not an issue, We define, $minQual((),(\texttt{hw "eth 10/100"})) = lo$, and so forth. However, if the switch needs to route all combinations of traffic, then a higher quality is required:

$$minQual((), (\texttt{hw ("ADSL" "eth 10/100" "eth 100/1000"}))) = hi$$

If we regard quality as reflecting the level of risk that a configuration must address, then there is a higher risk of failure when routing messages at multiple speeds versus simply routing at a single speed.   △

Quality Configuration. A node n is suitably configured if

$$minQual(minAttrib(n), maxAttrib(n)) \leq rating(n)$$

The above examples rely on a simple quality measure based on a total ordering $lo < med < hi$. We generalize this to a semiring [15]. Intuitively a semiring provides an addition operation on the set of measures. Some background information is provided in Appendix A.

5   The Configuration Cascade Problem

The quality of a network configuration is based not only on the quality of the configuration of the individual nodes, but is also based on their interoperation. This is illustrated by the following example.

Example 4. Consider a network composed of the nodes Bob and Clare, from Example 1. For simplicity, we assume that Bob hosts a kerberized mail service that allows limited classes of non-kerberized users to login based only on weak authentication (userid and password transmitted in clear text). Node Clare also provides a kerberized mail service and only supports strong authentication of non-kerberized clients; such clients must use SSH. Both of these nodes are deployed with *med* quality of configuration and are therefore acceptable by the configuration policy outlined in Example 2. Assume that the mail services hosted by Bob and Clare need to interoperate. We examine the quality of this network configuration.

An attacker has the ability to compromise systems that have *med* (or lower) quality of configuration. The attacker uses *dsniff* or *Ethereal* to obtain the email userid and password of a user of the service on Bob. Using a stack-smashing attack on the email service the attacker obtains copies of fresh Kerberos session keys and tickets. These are used to forge an authentic connection to the kerberized service on Clare and, in turn, some further vulnerability is exploited on Clare, allowing the attacker to access the SSH service.

The *med* quality nodes Bob and Clare individually meet the quality of configuration policy. By managing to compromise only *med* quality configurations this attacker can masquerade as a user who should be strongly authenticated (via SSH). However, the quality of the configuration policy requires that an attacker must compromise at least a *hi* quality configuration in order to misuse the attributes offered by the system:

$$minQual((),(\texttt{app "pop" "kerb" "ssh"})) = hi$$

When kerberos services on `Bob` and `Clare` are configured to interoperate a *cascading path* results in a violation of the the quality of configuration policy.   △

The traditional *cascade vulnerability problem* [17,18] is concerned with secure interoperation of compositions of multilevel secure systems that are evaluated to different levels of assurance according to the criteria specified in [17]. The transitivity of the multilevel security policy upheld across all secure systems ensures that their multilevel composition is secure; however, interoperability and data sharing between systems may increase the risk of compromise beyond that accepted by the assurance level.

We argue that cascades occur not just in multilevel secure systems, but in any network in which configuration quality is based on the quality of the individual nodes. The previous example illustrated this for a (non multilevel) security scenario. The next example illustrates this for a quality of service scenario.

Example 5. Consider a network composed of the node/switches `Eric` and `George`, as discussed in Example 3. We have,

$$rating(\texttt{Eric}) = med; \ rating(\texttt{George}) = med$$

and the nodes individually meet the quality of configuration policy.

Suppose that `Eric` and `George` are connected via a 10/100 Ethernet link. The effect of this configuration is that we have a sub-net that routes ADSL, 10/100 and 100/1000 Ethernet traffic. However, a higher quality is required:

$$minQual((), (\texttt{app ("ADSL" "eth 10/100" "eth 100/1000"))}) = hi$$

The failure of individual *med* quality configurations result in a risk that is supposed to be addressed by a *hi* quality configuration.   △

These two examples demonstrate that configuration quality can cascade when linking nodes together in a network. In practice, there are two issues to be addressed. Firstly, *detecting* cascading quality within a network requires evaluating all possible network paths to ensure that the required minimum quality is achieved over the attributes that are provided by all the nodes along the route. Secondly, if a cascade is discovered, then it is *eliminated* by either breaking appropriate links between nodes or replacing nodes with higher quality configurations until the cascade is eliminated.

The cascade problem has been previously studied in the context of multilevel secure systems. The model presented in this paper has a close relationship with the multilevel security model described in [19], whereby the lattice of multilevel security levels correspond to the lattice of attributes, the assurance levels correspond to the *Quality* c-semiring. Therefore, results on the multilevel cascade problem can be applied to our model. Existing research has considered schemes for detecting cascading multilevel security vulnerabilities and for eliminating them by reconfiguring system interoperation. While the detection of cascade vulnerabilities can be easily achieved [17,18] in polynomial time, their *optimal* elimination by breaking links is NP-complete [20].

The c-semiring representation also provides a convenient way to measure aggregate quality across collections of nodes that make up a network. The minimum quality of a series of nodes along a network path is given by the combination (under the c-semiring) of the quality ratings of the individual nodes. In this case, the weighted c-semiring $\mathcal{S}_{weight} = \langle \mathcal{R}, min, +, +\infty, 0 \rangle$ provides the appropriate measure. Given a series of possible paths that facilitate the interaction of x and y attributes then the effective quality of the path is the shortest path from x to y. There is a cascade vulnerability if the value calculated for this shortest path is more than $minQual(x, y)$. Practical techniques for calculating shortest paths across weighted constraint networks are considered in [15].

In suggesting the use of the weighted c-semiring $\mathcal{S}_{weight} = \langle \mathcal{R}, min, +, +\infty, 0 \rangle$ as one example of a quality measure, we are assuming that the risk of failure of one entity is independent of the risk of failure of any other entity. This is quite a restrictive assumption; however, there are examples where this kind of measure is useful. For example, in practice, the more firewalls/subnets that have to be traversed to directly access a node, then the more 'secure' the node is considered to be. The notion of *security distance* is defined in [21] as the minimum number of servers and/or firewalls that an attacker on the Internet must compromise to obtain direct access to some protected service. We argue that security distance in this case is equivalent to using a weighted c-semiring with each system having an equal rating of '1'. This can be generalized within our model to the *weighted security distance*, whereby, a weight is associated with each server and/or firewall to indicate the amount of effort that is required to compromise that component.

An alternative measure to using the weighted c-semiring is to interpret the probabilistic c-semiring $\mathcal{S}_{prob} = \langle \{x | \in [0, 1]\}, max, \times, 0, 1 \rangle$ [15] in terms of aggregation of risk along a path, which is calculated as combination (multiplication) of probabilities. As systems fail along a cascading path, then overall, there is increasing risk to subsequent systems failing.

6   Conclusion

In this paper we describe an abstract model of configuration that can be used to guide requirements for self-configuring nodes. The model takes a metric-based approach in that it allows networks to be built from nodes of varying quality.

A consequence of this model is that connecting correctly configured nodes together may result in a mis-configured system due to a cascading effect of their interoperation. These results have implications for the design of architectures that support self-configuring/autonomic nodes. When an autonomic node is introduced to a correctly configured network, it is not sufficient for the node to be just correctly configured; how the node interoperates with its neighbours can have a cascading impact on the overall quality of configuration. Therefore, each node must maintain and share sufficient information about the possible paths within the network configuration in order to able to determine whether it is safe to establish new connections. Doing this in an *optimal* way (for example, to maximise connectivity) may not be feasible since the cascade problem is, in general, hard.

Acknowledgements

We would like to thank the anonymous referees for their useful comments and feedback on the paper. This work has received partial support from the following: Enterprise Ireland (SC/2003/007); the MIUR Italian Project PRIN (n.2005-015491); Science Foundation Ireland (04/IN3/I404C and 00/PI.1/C075).

References

1. Drogseth, D., Hultquist, S., Nudler, J.: Network Performance Management: Three key technology challenges. Special Report, www.statseeker.com (2004)
2. Magrath, S., Chiang, F., Markovits, S., Braun, R., Cuervo, F.: Autonomics in Telecommunications Service Activation. First International Workshop on Autonomic Communication for Evolvable Next Generation Networks (2005)
3. Konstantinou, A., Florissi, D., Yemini, Y.: Towards Self-Configuring Networks. DARPA Active Networks Conference and Exposition (DANCE'02) (2002)
4. Ganek, A.G., Corbi, T.A.: The dawning of the autonomic computing era. IBM SYSTEMS JOURNAL, VOL 42, NO 1 (2003)
5. Horn, P.: Autonomic Computing: IBM's Perspective on the State of Information Technology. www.research.ibm.com/autonomic/manifesto (2001)
6. Kephart, J.O., Chess, D.M.: The Vision of Autonomic Computing. IEEE Computer Society (2003)
7. Balasubramaniam S., Barrett K., Strassner J., Donnelly W., van der Meer S.: Bio-inspired Policy Based Management (bioPBM) for Autonomic Communication Systems. 7th IEEE workshop on Policies for Distributed Systems and Networks, (2006)
8. TMF: TMF 053: The NGOSS Technology Neutral Architecture, (2005)
9. IBM: Policy Management for Autonomic Computing. IBM T.J. Watson Research Centre, (2005)
10. Durham D., et al: The COPS (Common Open Policy Service) Protocol. RFC 2748, (2000)
11. Westerinen A., Strassner J.: Common Information Model (CIM) Core Model. DSP0111, version 2.4, (2000)
12. Parker J.: FCAPS, TMN, ITIL: Three Key Ingerdients to Effictive IT Management. OpenWater Solutions, (2005)
13. Rivest, R.L.: S-expressions. Technical report, Network Working Group (1997) Internet Draft: http://theory.lcs.mit.edu/ rivest/sexp.txt.
14. Common Criteria Project: Common criteria for information technology security evaluation version 2.1. Technical report, US NIST (1999)
15. Bistarelli, S.: Semirings for Soft Constraint Solving and Programming. Volume LNCS 2962. Springer (2004)

16. Bistarelli, S., Montanari, U., Rossi, F.: Semiring-based Constraint Solving and Optimization. JACM **44**(2) (1997) 201–236 '
17. TNI: Trusted computer system evaluation criteria: Trusted Network Interpretation. Technical report, National Computer Security Center (1987) Red Book.
18. Millen, J., Schwartz, M.: The cascading problem for interconnected networks. In: 4th Aerospace Computer Security Applications Conference, IEEE CS Press (1988)
19. Foley, S.N., Bistaelli, S., O'Sullivan, B., Herbert, J., Swart, G.: Multilevel security and the quality of protection. In: Proceedings of First Workshop on Quality of Protection, Como, Italy, Springer Advances in Information Security, vol 23, 2006
20. Horton, R., et al.: The cascade vulnerability problem. Journal of Computer Security **2**(4) (1993) 279–290
21. Swart, G., Aziz, B., Foley, S., Herbert, J.: Trading off security in a service oriented architecture. In: 19th Annual IFIP WG 11.3 Working Conference on Data and Applications Security, (2005)

A   Appendix: Soft Constraints and c-Semirings

A semiring is a tuple $\langle \mathcal{S}, +, \times, \mathbf{0}, \mathbf{1} \rangle$ such that: $\mathcal{S}$ is a set and $\mathbf{0}, \mathbf{1} \in \mathcal{S}$; $+$ is commutative, associative and $\mathbf{0}$ is its unit element; $\times$ is associative, distributes over $+$, $\mathbf{1}$ is its unit element and $\mathbf{0}$ is its absorbing element. A c-semiring is a semiring $\langle \mathcal{S}, +, \times, \mathbf{0}, \mathbf{1} \rangle$ such that: $+$ is idempotent, $\mathbf{1}$ is its absorbing element and $\times$ is commutative.

Let us consider the relation $\leq_S$ over $\mathcal{S}$ such that $a \leq_S b$ iff $a + b = b$. Then it is possible to prove that (see [16]): $\leq_S$ is a partial order; $+$ and $\times$ are monotone on $\leq_S$; $\mathbf{0}$ is its minimum and $\mathbf{1}$ its maximum. Informally, the relation $\leq_S$ gives us a way to compare semiring values and constraints. In fact, when we have $a \leq_S b$, we will say that b *is better than* a. In the following, when the semiring will be clear from the context, $a \leq_S b$ will be often indicated by $a \leq b$.

The classical Constraint Satisfaction Problem (CSP) is a Soft CSP (SCSP) where the chosen c-semiring is: $S_{CSP} = \langle \{false, true\}, \vee, \wedge, false, true \rangle$. Fuzzy CSPs (FCSP) can instead be modelled in the SCSP framework by choosing the c-semiring $S_{FCSP} = \langle [0,1], max, min, 0, 1 \rangle$. Many other soft CSPs (probabilistic, weighted, ...) can be modelled by using a suitable semiring structure ($S_{prob} = \langle [0,1], max, \times, 0, 1 \rangle$, $S_{weight} = \langle \mathcal{R}, min, +, +\infty, 0 \rangle$, ...).

Risk Assessment of End-to-End Disconnection in IP Networks due to Network Failures *

Jens Milbrandt, Ruediger Martin, Michael Menth, and Florian Hoehn

University of Würzburg, Institute of Computer Science
Am Hubland, D-97074 Wurzburg, Germany
Phone: (+49) 931-888 6644; Fax: (+49) 931-888 6632
{milbrandt, martin, menth, hoehn}@informatik.uni-wuerzburg.de

Abstract. Restoration and protection switching mechanisms in IP networks are triggered by link or node failures to redirect traffic over backup paths. These mechanisms are no longer effective if a network becomes disconnected after a failure. The risk of end-to-end disconnection increases if the nodes of a network are only sparsely meshed or if multiple network failures occur simultaneously. This leads inevitably to violations of service level agreements with customers and peering network providers. In this paper, we present a method to assess the risk of end-to-end disconnection in IP networks due to network failures. We calculate the disconnection probabilities for all pairs of network nodes taking into account a set of probable network failures. The results are considered from different perspectives. This helps to identify weak spots of the network and to appropriately upgrade its topological infrastructure with additional links. We implemented the concept in a software tool which assists network providers with assessing the risk of disconnection in their network prior to any network failure and to take appropriate actions.

1   Introduction

Network resilience in carrier grade networks comprises the maintenance of both the connectivity and the quality of service (QoS) in terms of packet loss and delay during network failures. To maintain logical connectivity in the presence of link or node failures, restoration or protection switching mechanisms redirect traffic over backup paths. If the nodes of a network are only sparsely interconnected or if multiple network failures occur simultaneously, the risk of physical disconnectivity increases. As a consequence, resilience mechanisms are no longer effective if a network becomes disconnected after a failure. This leads inevitably to violations of service level agreements (SLAs) with customers or peering network providers. While the service availability in terms of maintainable logical connectivity is frequently in the focus of network analysis [1, 2, 3], the risk assessment of physical network paritioning due to network failures has not yet been investigated in depth.

   In this paper, we present a method to assess the risk of end-to-end (e2e) disconnection in IP networks due to network failures. Multiple independent or correlated network

---

* This work was funded by the Bavarian Ministry of Economic Affairs and the German Research Foundation (DFG). The authors alone are responsible for the content of the paper.

G. Parr, D. Malone, and M. Ó Foghlú (Eds.): IPOM 2006, LNCS 4268, pp. 181–192, 2006.

failures can occur. Correlated failures are due to shared risk resource groups (SR-RGs) [4]. For instance, links that are logically distinct on the network layer but share common resources on the link layer fail simultaneously if their common resource fails. We consider both types of multiple failures. The probability for multiple independent failures is rather low in a small network while in large networks their impact cannot be neglected. Our disconnection analysis therefore considers all relevant single and multi-failures that occur with a probability larger than a minimum threshold. Based on the probability of the relevant failure scenarios and their impact on the network connectivity, we calculate the disconnection probabilities for all pairs of nodes in the network. We consider the results for individual aggregates, individual routers, and for the overall traffic. It helps to validate SLAs, to get a quick overview on the overall network availability, to identify weak spots of the network and to appropriately upgrade its infrastructure with additional links. We implemented the concept in a software tool which helps network providers to assess the risk of e2e disconnection in their networks prior to any failure and to take appropriate actions.

This paper is structured as follows. In Section 2, we review related work regarding network resilience. Section 3 explains our algorithms to determine a set of relevant failure scenarios and to calculate thereon the aggregate-specific disconnection probabilities. Section 4 illustrates the analysis for an example network, suggests different views on the obtained data, and shows how our tool can help to upgrade the network topology. Finally, Section 5 concludes this work.

2   Network Failures and Resilience Mechanisms

In this section, we review causes for network failures. Some failures have only a local impact and the corresponding network outage can be compensated by resilience mechanisms. However, simultaneous multi-failures jeopardize the physical connectivity of a network and might cause its partitioning. We give an overview of current resilience mechanisms, discuss related work on network survivability, and comment on our contribution to assess the risk of physical network disconnection.

2.1   Network Failures

A good overview and characterization of causes for network failures is given in [5]. Basically, network failures with internal causes (e.g. software bugs, component defects, etc.) and can be distinguished from those with external causes (e.g. digging works, natural disaster, etc.). Furthermore, planned failure causes can be distinguished from unplanned failure causes. Planned outages are normally due to network maintenance and, since they are intentional, operators can take preventive measures. Planned network outages should not lead to physical network disconnections whereas unplanned outages jeopardize the network connectivity. The latter are difficult to predict and, therefore, operators must rely on resilience mechanisms in their networks. However, these mechanisms are no longer effective if a network becomes partitioned. To avoid physical disconnection, operators have to construct their network topology carefully.

Quantitative analyses and statistics about frequency and duration of failure events that occur in an operational network like the Sprint IP backbone are given in [6,7]. The authors detect all failures affecting the (logical) IP connectivity by analyzing the link state advertisements (LSAs) of interior gateway routing protocols. The results show that link failures are part of everyday's network operation and the majority of them is short-lived (less than 10 minutes). Moreover, they indicate that 20% of all failures are due to planned maintenance activities. Among the unplanned failures, almost 30% hit multiple links and can be attributed to router-related and optical equipment-related problems, while 70% affect only a single link at a time.

2.2 Resilience Mechanisms

Resilience mechanisms can be divided into restoration and protection schemes. Restoration sets up a new path after a failure while protection switching pre-establishes backup paths in advance. A good overview can be found in [5].

Usually, restoration is applied by IP rerouting. IP networks have the self-healing property, i.e., their routing re-converges after a network failure through the exchange of LSAs such that all but failed nodes can be reached after a while if the network is still physically connected. The reconvergence of the IP routing algorithm is a very simple and robust restoration mechanism [8, 9]. However, a clear disadvantage of IP routing is its slow convergence which is tolerable for elastic traffic but not for realtime traffic with strict QoS constraints. Another example for restoration besides IP rerouting are backup label switched paths (LSPs) in multi-protocol label switching (MPLS) that are set up after a network failure. However, such LSPs often follow the shortest paths of the interior gateway routing (IGP) protocol and, therefore, their setup requires even more time than pure IP rerouting.

Protection switching mechanisms address the problem of slow reconvergence speed. They can be implemented, e.g. in MPLS technology by explicitly routed and pre-established backup paths. Depending on the location of the reaction to a failure, protection switching mechanisms can be distinguished into e2e and local protection. In case of e2e protection switching, the reaction to a failure within a path is executed at the head end router. A simple concept simultaneously sets up (link or node disjoint) backup and primary paths between ingress/egress routers and in case of a failure traffic is just shifted from the broken primary path to the corresponding backup path. E2E protection switching is faster than restoration methods, but the signalling of the failure to the head end router of the path takes time within which traffic is lost. Local protection schemes tackle the problem of lost traffic in case of e2e protection. Backup paths are set up at the head end and at every intermediate router of the primary path. Therefore, a backup path is immediately available at any failure location along the primary path. Local protection switching is usually implemented on the physical or the data link layer due to the tight temporal constraints of effective network recovery (see [5]). However, recent developments also consider local protection switching on higher layers, e.g. MPLS fast reroute (MPLS-FRR) [10] or IP fast reroute mechanisms [11].

2.3 Network Survivability

The previosuly described resilience mechanisms are effective only if a network does not break apart after a failure, i.e., if all network nodes are still physically interconnected. Physical connectivity is a basic and most critical requirement for network survivability which can also be considered from different point of views [12]. Network survivability in terms of physical connectivity is a matter of topological network design and is thus subject to optimization [13]. Different network topologies have different survivability characteristics and require different strategies to improve survivability. Self-healing ring networks [14] represent a popular topology for metropolitan area networks (MANs) and have a good survivability potential since they operate with full hardware redundancy. Optical mesh networks [15, 16] characteristic for wide area networks (WANs) and their survivability can be improved by the installation of additional links. Network surviv-ability frameworks (cf. e.g. [12, 17]) define assessment and analysis models as well as different performance measures for the evaluation of multilayer network survivability. These frameworks are also applied in network analysis and network design [18, 19, 20].

2.4   Tools for Network Analysis and Network Design

We implemented our approach to assess the risk of e2e disconnection in IP network due to network failures in software since we are not aware of any standard tool for that purpose. Generally, standard simulation software like the OPNET Modeler [21] or the Network Simulator (ns-2) [22] provide means for analyzing the resilience of a network. However, these simulators focus on the dynamics of network resilience mechanisms with static sets of (single) network failures. They do not provide appropriate means for the calculation of probabilities for physical disconnection in a network for which a large amount of failures must be considered.

Other software products like e.g., the library of test instances for Survivable fixed telecommunication Network Design (SNDlib) [23], the TOolbox for Traffic Engineer-ing Methods (TOTEM) [24], or the NetScope tool for traffic engineering in IP net-works [25] focus on the evaluation of traffic engineering algorithms for network opti-mization. These tools cover only the usual network design and optimization problems such as routing, load balancing, flow allocation, and network dimensioning. However, they have no advanced functions to assess the risk of network disconnection.

2.5   Contribution of This Work

The above mentioned approaches are static in the sense that they respect only explicitly specified failures of (single) network elements. This is a reasonable start for resilient QoS provisioning, but the probability of multiple network failures grows with increas-ing network size. Therefore, simultaneous multi-failures need to be taken into account if the network size increases. Our objective is to assess the risk of e2e disconnection in IP networks due to network failures and to improve network survivability in terms of physical connectivity through identification of weak spots in the network.

The novelty of this work is the integration of outage probabilities in the assessment of physical survivability of a network. We present an assessment method that yields a complementary distribution function (CDF) for the e2e disconnection probability of the overall traffic and histograms for the e2e disconnection probabilities of routers and ag-gregates. This helps Internet service providers (1) to detect weak spots in their network and (2) to improve the survivability of their network by the systematical installation of additional links. We currently develop a tool that predicts the risk of e2e disconnec-tion before and after such a network modification to support the ISP in his decision process.

3 E2E Disconnection Due to Network Failures

We analyze the impact of potential failure scenarios on the physical connectivity of the network. As not all failure scenarios can be covered for that analysis, we determine only the most relevant. Some failure scenarios lead to the same working ("effective") topology which, in turn, leads to the same e2e disconnections in the network. We handle them jointly for the calculation of the disconnection probabilities.

3.1 Relevant Failure Scenarios

We first identify the relevant failure scenarios for our resilience analysis. To that aim, we collect all independent failure events $\hat{s} \in \hat{S}$. Note that this set may also contain shared risk groups (SRGs) such as shared risk link or node groups (SRLG, SRNG) [4]. Each of these failure events occurs with probability $p(\hat{s})$ and we number the events $\hat{s}_i$ in an descending order according to $p(\hat{s}_i)$. We define a compound failure scenario $s \subseteq \hat{S}$ as a subset of independent failure events that occur simultaneously with $p(s) = (\Pi_{\hat{s} \in s} p(\hat{s})) \cdot (\Pi_{\hat{s} \in \hat{S} \setminus s} (1 - p(\hat{s})))$. The set S contains all (compound) failure scenarios $s \subseteq \hat{S}$ with probability $p(s) \geq p_{min}$ where p_{min} is the probability threshold for relevant failure scenarios. Algorithm 1. constructs the set S starting with $S = \emptyset$. The recursive procedure is invoked with RELEVANTSCENARIOS$(0, \emptyset, 1)$. The algorithm steps recursively through the set of independent failure events $\hat{s}_i \in \hat{S}$. It constructs a compound failure scenario s incrementally and the recursion ends either if all $|\hat{S}|$ independent failure events $\hat{s}_i$ have been considered as potential members of s or if the probability $p(s)$ of the partial compound failure scenario s is too low[1]. In either case, scenario s joins S at the end of each recursion. At program termination, the set S contains all compound failure scenarios s that have a probability $p(s)$ larger than the threshold p_{min}.

Input: failure event number i, partial scenario s, and its probability $p(s)$
    if $(i = |\hat{S}|)$ **then**   {all independent failure events $\hat{s}_i$ have been considered}
        $S \leftarrow S \cup \{s\}$
    else if $(p(s) > p_{min})$ **then**   {partial scenario s is probable enough}
        RELEVANTSCENARIOS$(i+1, s \cup \hat{s}_i, p(s) \cdot p(\hat{s}_i))$
        RELEVANTSCENARIOS$(i+1, s, p(s) \cdot (1 - p(\hat{s}_i)))$
    end if

Algorithm 1. RELEVANTSCENARIOS: constructs the set of relevant scenarios S

3.2 Effective Topologies

The effective topology $T(s)$ caused by a compound failure scenario s is characterized by its set of working links and nodes. A link works only if itself and its adjacent routers do not fail. A router only works if itself and at least one of its adjacent links do not fail. Thus, all scenarios containing the failure of a router and some of its adjacent links lead to the same effective topology T. We subsume all of these scenarios in the set $S(T)$ and the probability of T is inherited by $p(T) = \sum_{s \in S(T)} p(s)$. The set $\mathcal{T} = \bigcup_{s \in S} T(s)$ denotes the set of all relevant effective topologies.

---

[1] The $|\mathcal{X}|$-operator denotes the cardinality of a set $\mathcal{X}$.

3.3   Calculation of Disconnection Probabilities

We denote the set of all nodes in a network by $\mathcal{V}$. The disconnection probability $p_{dis}^{\mathcal{S}}(v, w)$ of a single aggregate between two network nodes v and w is calculated by

$$p_{dis}^{\mathcal{S}}(v, w) = \frac{1}{p(\mathcal{S})} \cdot \sum_{s \in \mathcal{S}} p(s) \cdot \text{ISDISCONNECTEDIN}(v, w, T(s)) \qquad (1)$$

under the condition that only the relevant failure scenarios $s \in \mathcal{S}$ and their corresponding effective topologies $T(s)$ are respected. The function $\text{ISDISCONNECTEDIN}()$ implements a simple depth-first search or breadth-first search algorithm and yields either 1 if nodes v and w are disconnected in the topology T or 0 otherwise. Note that incremental shortest paths algorithms can speed up the computation time significantly [26]. The values $p_{dis}^{\mathcal{S}}(v, w)$ are underestimated since not all possible failure scenarios are considered in Equation (1). We can calculate an upper bound $p_{dis}^{max}(v, w)$ for the unconditioned disconnection probability by

$$p_{dis}^{max}(v, w) = p(\mathcal{S}) \cdot p_{dis}^{\mathcal{S}}(v, w) + (1 - p(\mathcal{S})) \cdot 1 \qquad (2)$$

under the assumption that the nodes v and w are disconnected in all unconsidered failure scenarios $s \in \hat{\mathcal{S}}^n \setminus \mathcal{S}$. Therefore, we rather use the conditional disconnection probabilities $p_{dis}^{\mathcal{S}}(v, w)$ to illustrate the application of our concept in Section 4 and compensate the inaccuracy by a very small threshold p_{min} such that the probability of covered failure scenarios is very close to 1.

4   Application of the Concept

This study is limited to link or node failures only. However, our software tool is able to handle correlated elemental failures of general shared risk resource groups as well.

4.1   Test Environment

To give a numerical example for our e2e disconnection analysis, we use the NOBEL network topology depicted in Figure 1. Set $\mathcal{V}$ comprises all network nodes and each of them is associated with a European city. For each pair of ingress/egress nodes v and w, we define a static aggregate rate

$$c(v, w) = \begin{cases} \dfrac{\pi(v) \cdot \pi(w) \cdot C}{\sum_{x, y \in \mathcal{V}, x \neq y} \pi(x) \cdot \pi(y)} & \text{if } v \neq w \\ 0 & \text{if } v = w \end{cases} \qquad (3)$$

where $\pi(v)$ is the population of city $v \in \mathcal{V}$ and C is the overall rate of all traffic aggregates. The populations $\pi(v)$ of all cities associated with the nodes $v \in \mathcal{V}$ are taken from [27]. They are shown in Figure 1 and are used to calculate the traffic matrix according to Equation (3).

The probability $p(\hat{s})$ of a failure event $\hat{s}$ depends on the availability A of the corresponding network element and it is defined by $p(\hat{s}) = 1 - A$. For the sake of simplicity, we set the node failure probabilities $p_{node} = 10^{-6}$ and compute the link failure probabilities p_{link} according to [5] as $p_{link} = \frac{\text{MTTR}}{\text{MTBF}}$ where MTTR = 12 h is the mean time

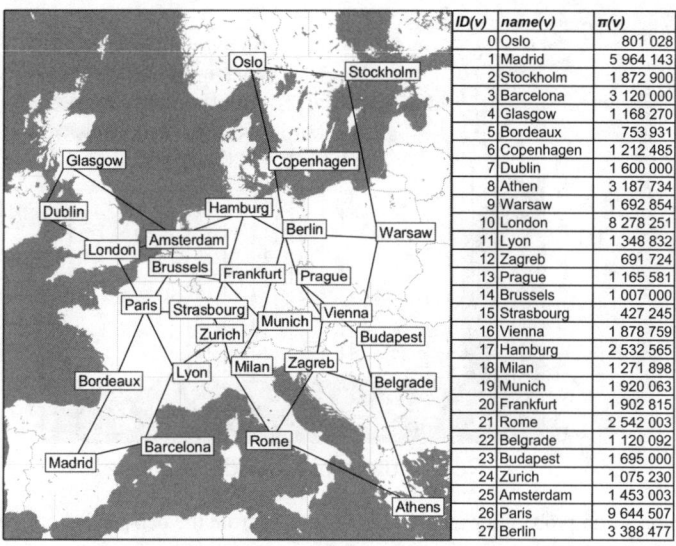

ID(v)	name(v)	π(v)
0	Oslo	801 028
1	Madrid	5 964 143
2	Stockholm	1 872 900
3	Barcelona	3 120 000
4	Glasgow	1 168 270
5	Bordeaux	753 931
6	Copenhagen	1 212 485
7	Dublin	1 600 000
8	Athen	3 187 734
9	Warsaw	1 692 854
10	London	8 278 251
11	Lyon	1 348 832
12	Zagreb	691 724
13	Prague	1 165 581
14	Brussels	1 007 000
15	Strasbourg	427 245
16	Vienna	1 878 759
17	Hamburg	2 532 565
18	Milan	1 271 898
19	Munich	1 920 063
20	Frankfurt	1 902 815
21	Rome	2 542 003
22	Belgrade	1 120 092
23	Budapest	1 695 000
24	Zurich	1 075 230
25	Amsterdam	1 453 003
26	Paris	9 644 507
27	Berlin	3 388 477

Fig. 1. European NOBEL test network and associated city populations

to repair and $\text{MTBF} = \frac{800\,\text{km}}{L} \cdot 365 \cdot 24$ h is the mean time between failures for a link l with length $L(l)$. We use a probability threshold of $p_{min} = 10^{-12}$ to calculate the relevant failure scenarios such that the remaining uncertainty is in the order of $1 - p(\mathcal{S}) \approx 10^{-6}$.

Disconnection Probabilities for Individual Aggregates. Figure 2 shows the disconnection probabilities of all aggregates from and to router Madrid to and from a single peer router w. Each column corresponds to a single bidirectional aggregate. The column width is proportional to the traffic volume $\frac{c(\text{Madrid},w)}{\sum_{v \in \mathcal{V}} c(\text{Madrid},v)}$ transported in both directions between router Madrid and its peer w and represents the lost traffic. The 27 peer routers are sorted along the x-axis in descending order of their disconnection probability in the original network. This perspective helps to verify the availability values for service level agreements (SLAs). If the e2e disconnection probabilities of specific aggregates are larger than allowed, either the SLA must be adjusted or the availability of these aggregates must be improved. They gray shaded columns in the figure correspond to such changes and are discussed later in this paper.

4.2   Disconnection Probabilities from the Perspective of the Overall Traffic

The e2e disconnection probability for source-destination pairs is calculated by Equation (1). Based on these values, we calculate the CDF of the e2e disconnection probability by

$$P(p_{dis}^{\mathcal{S}} > x) = \frac{1}{C} \cdot \sum_{\{v,w \in \mathcal{V}: p_{dis}^{\mathcal{S}}(v,w) > x\}} c(v,w). \tag{4}$$

under the condition that only the relevant failure scenarios $\mathcal{S}$ are respected. Figure 3 shows this CDF. The upmost curve corresponds to the original network while the others

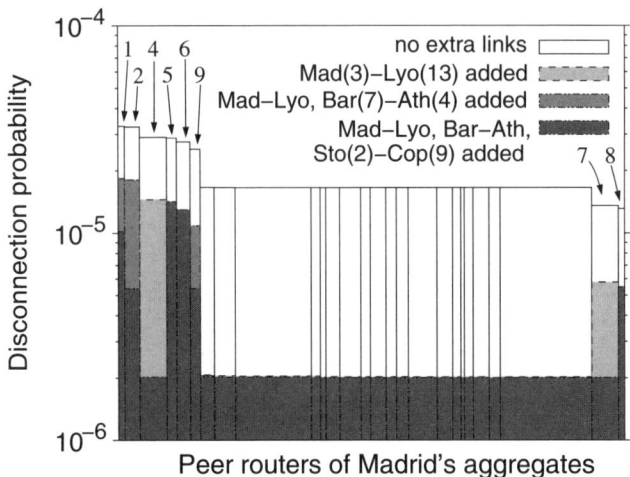

Fig. 2. Disconnection probabilities for traffic aggregates from the perspective of router *Madrid*

result from improved topologies that are discussed later. More than half of all aggregates are disconnected with a probability of $p_{dis}^S \in [10^{-5}, 3 \cdot 10^{-5}]$ while the probability for the remaining aggregates is one order of magnitude lower with $p_{dis}^S = 2 \cdot 10^{-6}$. This presentation provides a good overview regarding the overall availability of the network.

4.3 Disconnection Probabilities for Individual Routers

We calculate the average disconnection probabilities $p_{dis}^S(v)$ of the aggregates of individual routers by

$$p_{dis}^S(v) = \frac{\sum_{w \in \mathcal{V}, w \neq v} p_{dis}^S(v, w)}{(\mathcal{V} - 1)} + \frac{\sum_{w \in \mathcal{V}, w \neq v} p_{dis}^S(w, v)}{(\mathcal{V} - 1)} \tag{5}$$

under the condition that only the relevant failure scenarios S are respected.

Figure 4 shows p_{dis}^S for all 28 routers of the NOBEL network. The nodes are arranged along the x-axis according to their node ID from Figure 1 that have been issued in descending order of their disconnection probability in the original network for easier readability. Each column corresponds to a single router. The tall white columns represent the probabilities of the routers to get physically disconnected from other routers in the original network. The gray columns represent the disconnection probabilities after topology upgrades. Figure 4 shows that routers 1 through 9 have higher disconnection probabilities than the other nodes. They are connected to the network by only two links and are easily disconnected from the network by double link failures. Aggregates originating at or being destined to one of these nine nodes have also a larger disconnection probability which explains the large disconnection probability for large percentage of the overall traffic in Figure 3 and the large disconnection probabilities for some aggregates in Figure 2.

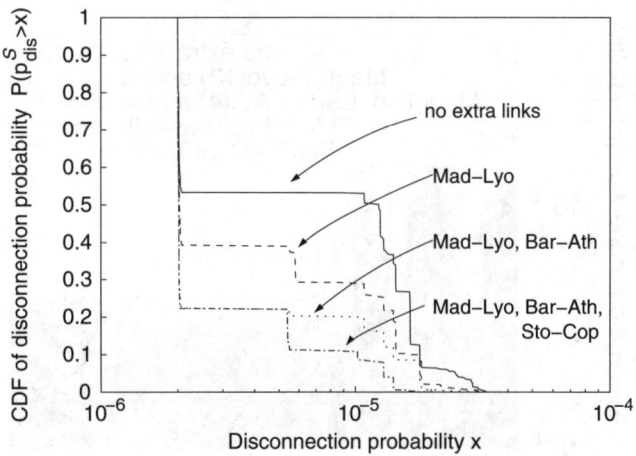

Fig. 3. Complementary distribution function of the e2e disconnection probability for the overall traffic

4.4 Improving the Network Availability

We have identified nodes with increased disconnection probabilities. The network operator can improve them, e.g., by installing additional links. For demonstration purposes, we first add the relatively short link Madrid-Lyon since Madrid (ID 3) is the biggest one among the nine critical nodes and therefore causes high traffic loss in case of disconnection from the network. As expected, Madrid profits most from this additional link, but also Barcelona (7) and Bordeaux (8) a higher gain in connectivity than the remaining routers in Figure 4. Figure 3 shows that this upgrade also improves the connectivity for the overall traffic significantly.

Next we add an additional link Barcelona-Athens since the two cities (7,4) are the major critical cities after Madrid. As a consequence, the disconnection probabilities decrease in Figures 3 and 4 due to this upgrade. The disconnection probabilities fall to $p_{dis}^S = 2 \cdot 10^{-6}$ for more traffic than before in Figure 3 because the added link improves the connectivity of two nodes. At $p_{dis}^S = 2 \cdot 10^{-6}$ the probabilities are dominated by the failure of either the source or the destination router of the respective aggregates. Thus, the disconnection probability cannot be reduced below $2 \cdot 10^{-6}$ by adding additional links. However, increasing the node reliability, e.g. by installing redundant hardware, or multi-homing can further improve the network availability.

We select the third additional link Stockholm-Copenhagen to improve the availability of both medium size cities (2,9) since they are close to each other. However, Figures 3 shows only improvements to larger e2e disconnection probabilities ($5.5 \cdot 10^{-6}$) but not to $2 \cdot 10^{-6}$. Likewise, the e2e disconnection probability of the nodes with a large unavailability cannot reduced to a low value by the introduction of the new link (cf. Figure 4). This is due to the double link failures Berlin-Copenhagen and Warsaw-Stockholm that can disconnect the triangle Oslo-Stockholm-Copenhagen completely from the network.

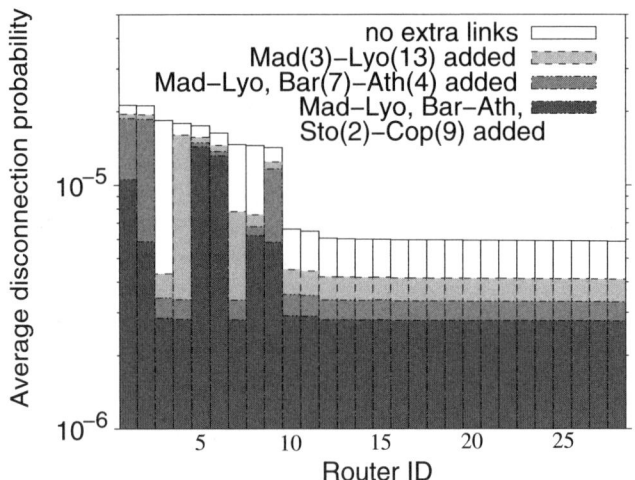

Fig. 4. Disconnection probabilities for individual routers

The assessment and visualization of the e2e disconnection probability enables network operators to identify weak spots in the network such that they can apply topology improvements and evaluate their impact. Our tool confirms intuition and gives additional hints since it considers much more possible scenarios than can be seen at first sight and supports network planners in strategic decisions.

5   Conclusion

Resilience mechanisms maintain the connectivity of a network in case of failures, but this is not always possible, e.g., in the presence of severe multi-failures. In this paper, we proposed a method to assess the risk of end-to-end (e2e) disconnection in IP networks taking into account all failures with a probability larger than a threshold p_{min}.

We illustrated this concept in the European Nobel network and used IP rerouting as resilience mechanism. The e2e disconnection probability for individual aggregates helps to define suitable service level agreements. The complementary distribution function of the e2e disconnection probability for the overall traffic provides a good overview of the connectivity of the entire network. The e2e disconnection probability for individual routers helps to detect weakly connected routers, to upgrade the network by additional links, and to visualize their impact.

We implemented the concept in a software tool which supports network providers to assess the risk of e2e disconnection in their networks prior to any network failure and to take appropriate actions in advance. Currently, we extend our tool towards the a priori detection of overload [28] caused by rerouted traffic and traffic variations that may occur due to inter-domain rerouting [29] or due to traffic hot spots [30].

References

1. Chandra, B., Dahlin, M., Gao, L., A.Nayate: End-to-End WAN Service Availability. In: Proc. of 3rd USENIX Symposium on Internet Technology & Systems (USITS), San Francisco, USA (2001) 97–108
2. Bhattacharyya, S., Diot, C., Iannaccone, G., Markopoulou, A., Chuah, C.: Service Availability in IP Networks. ATL Research Report RR03-ATL-071888, Sprint (2003)
3. Keralapura, R., Chuah, C., Iannaconne, G., Bhattacharrya, S.: Service Availability: A New Approach to Characterize IP-Backbone Topologies. In: Proc. of 12th International Workshop on Quality of Service (IWQoS), Montreal, Canada (2004) 232–241
4. Datta, P., Somani, A.K.: Diverse Routing for Shared Risk Resource Groups (SRRG's) in WDM Optical Networks. In: 1st International Conference on Broadband Communication, Networks, and Systems (BROADNETS). (2004) 120 – 129
5. Vasseur, J., Pickavet, M., Demeester, P.: Network Recovery. Elsevier (2004)
6. Iannaccone, G., Chuah, C., Mortier, R., Bhattacharyya, S., Diot, C.: Analysis of Link Failures in an IP Backbone. In: Proc. of ACM SIGCOMM Internet Measurement Workshop, Marseille, France (2002) 237–242
7. Markopoulou, A., Iannaccone, G., Bhattacharyya, S., Chuah, C., Diot, C.: Characterization of Failures in an IP Backbone. In: Proc. of 23rd IEEE Infocom, Hong Kong, China (2004)
8. Nucci, A., Schroeder, B., Bhattacharyya, S., Taft, N., Diot, C.: IGP Link Weight Assignment for Transient Link Failures. In: 18th International Teletraffic Congress (ITC), Berlin (2003)
9. Fortz, B., Thorup, M.: Robust Optimization of OSPF/IS-IS Weights. In: International Network Optimization Conference (INOC), Paris, France (2003) 225–230
10. Pan, P., Swallow, G., Atlas, A.: RFC4090: Fast Reroute Extensions to RSVP-TE for LSP Tunnels (2005)
11. Shand, M., Bryant, S.: IP Fast Reroute Framework. http://www.ietf.org/internet-drafts/draft-ietf-rtgwg-ipfrr-framework-05.txt (2006)
12. Zolfaghari, A., Kaudel, F.J.: Framework for Network Survivability Performance. IEEE Journal on Selected Areas in Communications **12**(1) (1994) 46–51
13. Boorstyn, R.R., Frank, H.: Large-Scale Network Topological Optimization. IEEE Transactions on Communications **25**(1) (1977) 29–47
14. Towster, H., Stephenson, R.W., Morgan, S., Keller, M., Mayer, R., Shalayda, R.: Self-Healing Ring Networks: Gateway to Public Information Networking. IEEE Communications Magazine **28**(6) (1990) 54–60
15. Ramamurthy, S., Mukherjee, B.: Survivable WDM Mesh Networks, Part I - Protection. In: Proc. of IEEE Conference on Computer Communications (INFOCOM). (1999) 744 – 751
16. Ramamurthy, S., Mukherjee, B.: Survivable WDM Mesh Networks, Part II - Restoration. In: Proc. of IEEE International Conference on Communications (ICC). (1999) 2023 – 2030
17. Medhi, D.: A Unified Approach to Network Survivability for Teletraffic Networks: Models, Algorithms and Analysis. IEEE Transactions on Communications **42**(2) (1994) 534–548
18. Cardwell, R.H., Monma, C.L., Wu, T.H.: Computer-Aided Design Procedures for Survivable Fiber Optic Networks. IEEE Journal on Selected Areas in Communications **7**(8) (1989) 1188–1197
19. Gavish, B., Trudeau, P., Dror, M., Gendreau, M., Mason, L.: Fiberoptic Circuit Network Design under Reliability Constraints. IEEE Journal on Selected Areas in Communications **7**(8) (1989) 1181–1187
20. Cankaya, H., Lardies, A., Ester, G.: Network Design Optimization from an Availability Pespective. In: International Telecommunication Network Strategy and Planning Symposium (Networks), Vienna, Austria (2004)

21. OPNET Incorporated: OPNET Modeler. `http://opnet.com/products/modeler/` (1987)
22. University of Southern California - Information Sciences Institute (USC-ISI): Network Simulator (ns-2). `http://www.isi.edu/nsnam/ns/` (1995)
23. Zuse-Institute Berlin (ZIB):   SNDlib 1.0 – Survivable Network Design Data Library. `http://sndlib.zib.de` (2005)
24. Balon, S., Lepropre, J., Monfort, G.: TOTEM 2.0 - TOolbox for Traffic Engineering Methods. `http://totem.run.montefiore.ulg.ac.be` (2005)
25. Feldmann, A., Greenberg, A., Lund, C., Reingold, N., Rexford, J.: NetScope: Traffic Engineering for IP Networks. IEEE Network Magazine (2000) 11–19
26. Menth, M., Milbrandt, J., Lehrieder, F.: Algorithms for Fast Resilience Analysis in IP Networks. In: 6th IEEE Workshop on IP Operations and Management (IPOM), Dublin, Ireland (2006)
27. Brinkhoff, T.:   Population of the Major Cities and Agglomerations for Each Country. http://www.citypopulation.de/ (1998-2006)
28. Menth, M., Milbrandt, J., Hoehn, F., Lehrieder, F.: Assessment and Visualization of E2E Unavailability and Potential Overload in Communication Networks. In: currently under submission. (2006)
29. Schwabe, T., Gruber, C.G.: Traffic Variations Caused by Inter-domain Re-routing. In: International Workshop on the Design of Reliable Communication Networks (DRCN), Ischia Island, Italy (2005)
30. Menth, M., Martin, R., Charzinski, J.: Capacity Overprovisioning for Networks with Resilience Requirements. In: ACM SIGCOMM, Pisa, Italy (2006)

Evaluation of a Large-Scale Topology Discovery Algorithm

Benoit Donnet[1,2], Bradley Huffaker[2], Timur Friedman[1], and kc claffy[2]

[1] Université Pierre & Marie Curie – Laboratoire LiP6/CNRS, UMR 7606, France
[2] CAIDA – San Diego Supercomputer Center, USA

Abstract. In the past few years, the network measurement community has been interested in the problem of internet topology discovery using a large number (hundreds or thousands) of measurement monitors. The standard way to obtain information about the internet topology is to use the traceroute tool from a small number of monitors. Recent papers have made the case that increasing the number of monitors will give a more accurate view of the topology. However, scaling up the number of monitors is not a trivial process. Duplication of effort close to the monitors wastes time by reexploring well-known parts of the network, and close to destinations might appear to be a distributed denial-of-service (DDoS) attack as the probes converge from a set of sources towards a given destination. In prior work, authors of this paper proposed Doubletree, an algorithm for cooperative topology discovery, that reduces the load on the network, i.e., router IP interfaces and end-hosts, while discovering almost as many nodes and links as standard approaches based on traceroute. This paper presents our open-source and freely downloadable implementation of Doubletree in a tool we call traceroute@home. We evaluate the performance of our implementation on the PlanetLab testbed and discuss a large-scale monitoring infrastructure that could benefit of Doubletree.

1 Introduction

Today's most extensive tracing system at the IP interface level, *skitter* [1], uses 24 monitors, each targeting on the order of one million destinations. Authors of this paper are responsible for skitter. In the fashion of skitter, *scamper* [2] makes use of several monitors to traceroute IPv6 networks. Other well known systems, such as RIPE *NCC TTM* [3] and NLANR *AMP* [4], each employ a larger set of monitors, on the order of one- to two-hundred, but they avoid probing outside their own network. However, recent work has indicated the need to increase the number of traceroute sources in order to obtain a more complete topology measurement [5,6]. Indeed, it has been shown that reliance upon a relatively small number of monitors to generate a graph of the internet can introduce unwanted biases.

One way of rapidly creating a large distributed monitor infrastructure would be to deploy traceroute monitors in an easily downloadable and readily usable

G. Parr, D. Malone, and M. Ó Foghlú (Eds.): IPOM 2006, LNCS 4268, pp. 193–204, 2006.

piece of software, such as a screensaver. This was first proposed by Jörg Non-nenmacher, as reported by Cheswick et al. [7]. Such a suggestion is in keeping with the spirit of that have arisen in the past few years. The most famous one is probably SETI@home [8]. SETI@home's screensaver downloads and analyzes radio-telescope data. The first publicly downloadable distributed route tracing tool is DIMES [9], released as a daemon in September 2004. At the time of writing this paper, DIMES counts more than 8,700 agents scattered over five continents.

However, building such a large structure leads to potential scaling issues: the quantity of probes launched might consume undue network resources and the probes sent from many vantage points might appear as a distributed denial-of-service (DDoS) attack to end-hosts. These problems were quantified in our prior work [10].

The Doubletree algorithm [10], proposed by two authors of this paper, is an attempt to perform large-scale topology discovery efficiently and in a network friendly manner. Doubletree acts to avoid retracing the same routes in the internet by taking advantage of the tree-like structure of routes fanning out from a source or converging on a destination. The key to Doubletree is that monitors share information regarding the paths that they have explored. If one monitor has already probed a given path to a destination then another monitor should avoid that path. Probing in this manner can significantly reduce load on routers and destinations while maintaining high node and link coverage [10]. By avoiding redundancy, not only is Doubletree able to reduce the load on the network but it also allows one to probe the network more frequently. This makes it possible to better capture network dynamic (routing changes, load balancing) compared to standard approaches based on traceroute.

This paper goes beyond earlier theory and simulation to propose a Doubletree implementation in tool called *traceroute@home*. traceroute@home is open-source and freely available [11][1]. The goal of this paper is neither to compare Doubletree to standard probing (e.g., skitter) in a real environment, neither to tell the story of the large-scale deployment of Doubletree. We aim to evaluate our implementation of the algorithm and understand its behavior in a real, but controlled, environment, i.e., PlanetLab [13]. This approach can be seen as a first step towards a larger-scale deployment of the algorithm in an entirely dedicated measurement structure. We first implement and evaluate the core of the measurement system, i.e., the probing engine, before building a more complex infrastructure. Our implementation is modular, making future extensions and reuse in a dedicated measurement structure easy. In this paper, we also discuss a large-scale measurement infrastructure that could benefit of traceroute@home.

The remainder of this paper is organized as follows: Sec. 2 describes the Doubletree algorithm; Sec. 3 describes traceroute@home; in Sec. 4, we discuss the performance evaluation done on PlanetLab nodes; Sec. 5 discusses the usage of traceroute@home in an entirely dedicated measurement infrastructure; finally, Sec. 6 summarizes the principal contributions of this paper.

---

[1] Interested readers might find an extended version of this paper in an arXiv technical report [12].

2   Doubletree

Doubletree [10] is the key component of a coordinated probing system that significantly reduces load on routers and end-hosts while discovering nearly the same set of nodes and links as standard approaches based on traceroute. It takes advantage of the tree-like structures of routes in the context of probing. Routes leading out from a monitor towards multiple destinations form a tree-like structure rooted at the monitor. Similarly, routes converging towards a destination from multiple monitors form a tree-like structure, but rooted at the destination. A monitor probes hop by hop so long as it encounters previously unknown interfaces. However, once it encounters a known interface, it stops, assuming that it has touched a tree and the rest of the path to the root is also known. Using these trees suggests two different probing schemes: backwards (monitor-rooted tree) and forwards (destination-rooted tree).

For both backwards and forwards probing, Doubletree uses stop sets. The one for backwards probing, called the *local stop set*, consists of all interfaces already seen by that monitor. Forwards probing uses the *global stop set* of (interface, destination) pairs accumulated from all monitors. A pair enters the stop set if a monitor received a packet from the interface in reply to a probe sent towards the destination address.

A monitor that implements Doubletree starts probing for a destination at some number of hops h from itself. It will probe forwards at $h + 1$, $h + 2$, etc., adding to the global stop set at each hop, until it encounters either the destination or a member of the global stop set. It will then probe backwards at $h - 1$, $h - 2$, etc., adding to both the local and global stop sets at each hop, until it either has reached a distance of one hop or it encounters a member of the local stop set. It then proceeds to probe for the next destination. When it has completed probing for all destinations, the global stop set is communicated to the next monitor.

Doubletree has one tunable parameter. The choice of initial probing distance h is crucial. Too close, and duplication of effort will approach the high levels seen by classic forwards probing techniques [10, Sec. 2]. Too far, and there will be high risk of traffic looking like a DDoS attack for destinations. The choice must be guided primarily by this latter consideration to avoid having probing look like a DDoS attack.

While Doubletree largely limits redundancy on destinations once hop-by-hop probing is underway, its global stop set cannot prevent the initial probe from reaching a destination if h is set too high. Therefore, each monitor sets its own value for h in terms of the probability p that a probe sent h hops towards a randomly selected destination will actually hit that destination. Fig. 2 shows the cumulative mass function for this probability for skitter monitor `champagne`. If one considers as reasonable a 0.2 probability of hitting a responding destination on the first probe, it must chose $h \leq 14$.

Simulation results [10, Sec. 3.2] show for a range of p values that, compared to classic probing, Doubletree is able to reduce measurement load by approximately 70% while maintaining interface and link coverage above 90%.

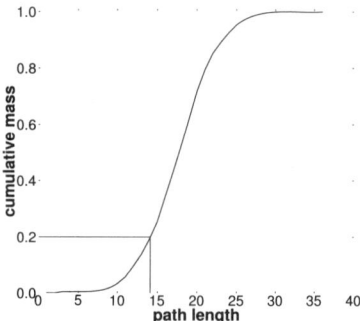

Fig. 1. Cumulative mass plot of path lengths from skitter monitor `champagne`

Doubletree assumes, in the context of probing, that the routes have a tree-like structure. This is true in a large proportion, as suggested by Doubletree's coverage results (see [10, Sec. 3.2]), but this hypothesis implies a static view of the network. When a Doubletree monitor stops probing towards the root of a tree, it is making the bet that the rest of the path to the tree is both know and unchanged since earlier probing. The existence of routes' convergence and divergence points, however, imply a dynamic view of the network, as some parts of the network might change due to load balancing and routing. We are currently working on improving Doubletree in order to take into account dynamic behaviors of the network [14].

3   Implementation

In this section, we describe our implementation of the Doubletree algorithm in a tool called traceroute@home. We first introduce our design choices (Sec. 3.1) and, next, we give an overview of the system (Sec. 3.2).

3.1   Design Choices

We implemented traceroute@home in Java [15]. We choose Java as the development language because of two reasons: the large quantity of available packages and the possibility of abstracting ourselves from technical details. As a consequence, the development time was strongly reduced. Unfortunately, Sun does not provide any package for accessing packet headers and handling raw sockets, which is necessary to implement traceroute. Instead of developing our own raw sockets library, we used the open-source *JSocket Wrench* library [16]. We modified the JSocket Wrench library in order to support multi-threading. Our modifications are freely available [11].

We aimed for the design of traceroute@home to be easily extended in the future by ourselves but also by the networking community. For instance, concerning

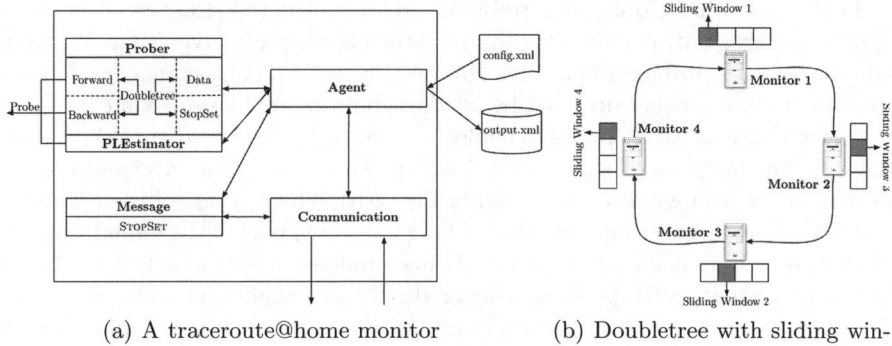

(a) A traceroute@home monitor

(b) Doubletree with sliding window

Fig. 2. The traceroute@home system

the messages exchanged by monitors, we define a general framework for messages, making creation and handling of new messages easier.

We designed our application by considering two levels: the *microscopic* level and the *macroscopic* level.

From a macroscopic point of view, i.e., all the monitors together, the monitors are organized in a ring, adopting a round robin process. At a given time, each monitor focuses on its own part of the destination list. When it finishes probing its part, it sends information to the next monitor and waits for data from the previous one, if it was not yet received. Sec. 3.2 explains this macroscopic aspect of traceroute@home.

From a microscopic point of view of our implementation, i.e., a single monitor, a monitor is tuned with an XML configuration file loaded at its starting. A traceroute@home monitor is composed of several modules that interact with each other. Our implementation is thread-safe, as a monitor, conducted by the *Agent* module, is able to send several probes (ICMP or UDP) and receive several messages from other monitors at the same time. Further, topological information collected by a monitor is regularly saved to XML files. Obviously, a traceroute@home monitor implements a *Doubletree* module, as described in Sec. 2. Fig. 2(a) illustrates a traceroute@home monitor. Going into deeper details within a traceroute@home monitor is beyond the scope of this paper. Nevertheless, interested readers might find a complete description of a traceroute@home monitor in [12].

Our implementation is open-source and freely available [11].

3.2 System Overview

The simulations conducted in prior work [10] were based on a simple probing system: each monitor in turn covers the destination list, adds to the global stop set the (interface, destination) pairs that it encounters, and passes the set to the subsequent monitor.

This simple scenario is not suitable in practice: it is too slow, as an iterative approach allows only one monitor to probe the network at a given time. We want all the monitors probing in parallel. However, how would one manage the global stop set if it were being updated by all the monitors at the same time?

An easy way to parallelize is to deploy several *sliding windows* that slide along the different portions of the destination list. At a given time, a given monitor focuses on its own window, as shown in Fig. 2(b). There is no collision between monitors, in the sense that each one is filling in its own part of the global stop set. The entire system counts m different sliding windows, where m is the number of Doubletree monitors. If there are n destinations, each window is of size $w = n/m$. This is an upper-bound on the window size as the concept still applies if they are smaller.

A sliding window mechanism requires us to decide on a step size by which to advance the window. We could use a step size of a single destination. After probing that destination, a Doubletree monitor sends a small set of pairs corresponding to that destination to the next monitor, as its contribution to the global stop set. It advances its window past this destination, and proceeds to the next destination. Clearly, though, a step size of one will be costly in terms of communication. Packet headers (see [12] for details about packet format) will not be amortized over a large payload, and the payload itself, consisting of a small set, will not be as susceptible to compression as a larger set would be.

On the other hand, a step size equal to the size of the window itself poses other risks. Suppose a monitor has completed probing each destination in its window, and has sent the resulting subset of the global stop set on to the following monitor. It then might be in a situation where it must wait for the prior monitor to terminate its window before it can do any further useful work.

A compromise must be reached, between lowering communications costs and continuously supplying each monitor with useful work. This implies a step size somewhere between 1 and w. For our implementation of Doubletree, we let the user decide the step size. This is a part of the XML configuration file that each Doubletree monitor loads at its start-up (see Fig. 2(a) and [12]). Future work might reveal information about how to tune the step size of a monitor.

4   Performance Evaluation

As described in prior work [10], security concerns are paramount in large-scale active probing. It is important to not trigger alarms inside the network with Doubletree probes. It is also important to avoid burdening the network and the destination hosts. It follows from this that the deployment of a cooperative active probing tool must be done carefully, proceeding step by step, from an initial small size, up to larger-scales. Note that this behavior is strongly recommended by PlanetLab [17, Pg. 5].

We tested traceroute@home on a set of ten PlanetLab nodes. These ten nodes acted as traceroute@home monitors. We selected them based on their relatively high stability (i.e., remaining up and connected), and their relatively low load.

Table 1. Stopping reasons (in %) and h value per monitor

monitor	Backwards				Forwards				h
	loop	gap	stop set	normal	loop	gap	stop set	normal	
Blast	0	0	99.5	0.5	2	17	50	31	7
Cornell	0	0	99	1	0	13.5	69.5	17	7
Ethz	1	0	98.5	0.5	2	10.5	52	35.5	11
Inria	1.5	0	97.5	1	1	4	67	28	15
Kaist	0	0	99	1	0.5	10.5	64.5	24.5	9
Nbgisp	0.5	4	95	0.5	3.5	30.5	22	44	7
LiP6	0	0	99.5	0.5	1	9.5	62.5	27	11
UCSD	0	0	99.5	0.5	0	10.5	60.5	29	7
Uoregon	0	0	99.5	0.5	0	7	74.5	18.5	6
Upc	0.5	0	99	0.5	1	14	57.5	27.5	15
mean	0.35	0.4	98.6	0.65	0.11	12.7	58	28.2	9

These traceroute@home monitors are scattered around the world: North America (USA, Canada), Europe (France, Spain, Switzerland, Spain), and Asia (Japan, Korea). Evaluating the performance of selected PlanetLab nodes is beyond the scope of this paper. Interested readers might find further information about such an evaluation in [12].

The destination list, i.e., the probe targets consists of $n = 200$ PlanetLab nodes randomly chosen amongst the approximately 300 institutions that currently host PlanetLab nodes. Restricting ourselves to PlanetLab nodes destinations was motivated by security concerns. By avoiding tracing outside the PlanetLab network, we avoid disturbing end-systems that do not welcome probe traffic. None of the ten PlanetLab monitors (or other nodes located at the same place) belongs to this destination list. The sliding window size of $w = n/m$ consists of twenty destinations. We consider two step sizes by window, so each step counts ten destinations. Each traceroute@home monitor was configured as follows: the probability p was set to 0.05, no compression was required before sending messages and a sliding window was composed of two step sizes.

The experiment was run on the PlanetLab nodes on Dec. 20th 2005. All the traceroute@home monitors were started at the same time. The experiment was finished when each monitor had probed the entire destination list.

A total of 2,703 links and 2,232 nodes were discovered. We also encountered 2,434 non-responding interfaces (routers and destinations). We recorded 36 invalid addresses. Invalid addresses are, for example, private addresses [10, Sec. 2.1].

Table 1 shows the different reasons for stopping backwards and forwards probing for each traceroute@home monitor. It further indicates the h value computed by each monitor. The last row of the table indicates the mean for each column. A *loop* occurs when a given node appears at two different hops. A *gap* occurs when five successive nodes does not reply to probes. A *stop set* indicates the application of a stopping rule based on the membership to a given stop set (local stop set for backwards and global stop set for forwards), as defined in Sec. 2). A *normal* stopping means hitting the first hop (backwards) or the destination (forwards).

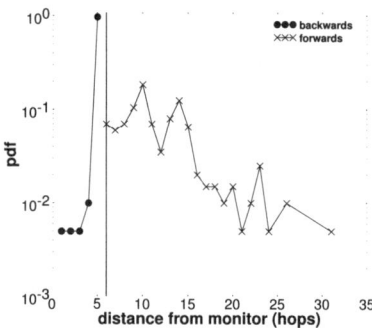

Fig. 3. Stopping distance for the Uoregon monitor

Looking first at the backwards stopping reasons, we see that the stop set rule strongly dominates (98.6% on average). On average, normal stopping occurs only 0.65% of the time.

Fig. 3 shows the stopping distance (in terms of hops from the monitor), for the Uoregon monitor, when probing backwards and forwards. The vertical line indicates the h value computed by Uoregon. Results presented in Fig. 3 are typical for the other traceroute@home monitors.

We see that more than 90% of the backwards stopping occurs at a distance of 5, that is to say the distance corresponding to $h - 1$. In 2.5% of the cases, the probing stops between hop 1 (normal stopping) and hop 4. Except for hop 1, the other stops might be caused by the stop set or by hitting a destination, probably due to very short paths. This latter case illustrates a situation in which the first probe sent with a TTL of h directly hits a destination.

Looking now at the forwards stopping reasons in Table 1, we see that the gap rule plays a greater rule. We believe that these gaps occur when a destination does not respond to probes because of a restrictive firewall or because the PlanetLab node is down.

On average, in 58% of the cases, the stop set rule applies, and in 28.2% of the cases, the normal rule applies. The normal rule proportion might be seen as high but we have to keep in mind that a Doubletree monitor starts with an empty stop set. Therefore, during the first sliding window, the only thing that can stop a monitor, aside from the gap rule, is an encounter with the destination.

Looking at the stopping distance in Fig. 3, we see that the distances are more scattered for forwards probing than for backwards probing. Regarding the forwards probing, a peak is reached at a distance of 10 (18.5% of the cases). In 7% of the cases, the monitor stops probing at a distance of 6, that is equal to the value h. It could correspond to the stop set rule application or the normal rule, by definition of p. Recall that p defines the probability of hitting a destination with the probe sent with a TTL equals to h. For our experiment, we set $p = 0.05$, meaning that in 5% of the cases the first probe sent by a monitor will hit a destination.

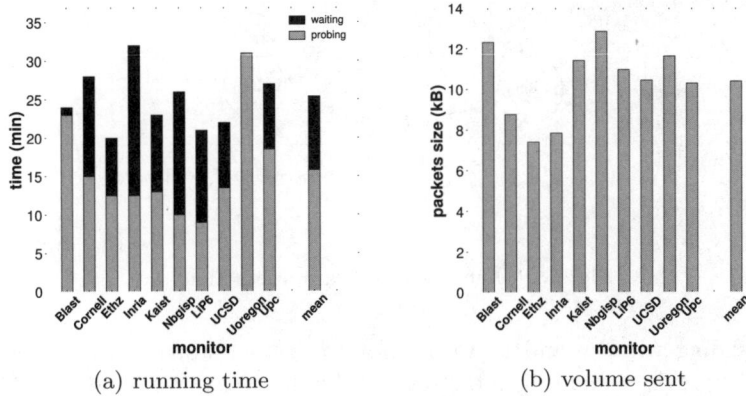

(a) running time   (b) volume sent

Fig. 4. Load

Fig. 4 shows, for each traceroute@home monitor, the load generated by our prototype. This load is expressed in terms of the running time (Fig. 4(a)) and of the total size of packets exchanged by monitors (Fig. 4(b)). Each figure has an additional bar on the right of the plot that gives the mean over the ten monitors.

The size of packets takes into account the header (4 bytes) and the payload. Interested readers might find further information about messages in [12]. The messages exchanged by monitors are STOPSET messages. A STOPSET message is sent by a monitor when it reaches a step size in the current sliding window and contains stop set information for the next monitor in the ring (See Fig. 2(b)). As we define for our experiment two step sizes per sliding window and as we deploy our prototype on ten PlanetLab nodes, each monitor sent 20 STOPSET messages.

The monitors do not exchange their entire stop set. They only send an update that contains the (interface, destination) pairs discovered during the current step size probing.

In Fig. 4(b), we can see that a monitor sends a total of between 7.41 KB and 12.84 KB to the subsequent monitor. On average, a monitor sends 10.39 KB of stop set information into the network.

During our experimentation, the traceroute@home application did not flood the network with STOPSET messages. However, our prior work [18] has shown, on a larger destination list, that it can grow to excessive sizes. In this case, our prior work suggests to implement the global stop set as a Bloom filter [18] instead of a list of (interface, destination) pairs. This implementation is provided in our prototype. Our prototype is easily tunable, due to the use of an XML configuration file. The user must specify in this file which type of implementation the prototype has to use. For our experiments, we choose to consider the standard implementation of the global stop set, i.e., the list.

Looking now at the running time (Fig. 4(a)), we see that it is expressed as a combination of probing (gray bars) and waiting periods (black bars). The waiting period occurs when a monitor has finished its sliding window or a step size in

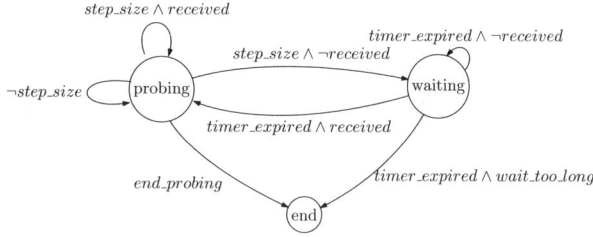

Fig. 5. Probing/waiting state interactions

a given sliding window and is waiting for the global stop set that should be sent by the previous monitor in the round-robin topology. We see that nearly all monitors have to wait. A waiting period, in our implementation, lasts 30 seconds. When the timer expires, the monitor checks if it received a new message. If so, the waiting period ends and a new probing period begins. Otherwise, it sleeps during 30 seconds. To avoid infinite waiting, if after 40 sleeping periods (i.e., 20 minutes), nothing was received, the monitor quits with an error message. Fig. 5 illustrates the interactions between the probing state and the waiting state.

We believe that these long waiting periods are due to a characteristic of the PlanetLab IP stack. It seems that when ICMP replies are by the stack, the socket reader function does not read them immediately. As the timer set on the listening socket never expires in this case, we think that the socket reader function is waiting for the permission to access the IP stack. It looks like the resource is owned (or locked) by another process on the PlanetLab node. Note that this behavior was also noticed by other Planet-Lab users [19].

5 Measurement Infrastructures

The recent NSF-sponsored CONMI Workshop [20] (in which two of the present authors participated) urged a comprehensive approach to distributed probing, with a shared infrastructure that respects the many security concerns that active measurements raise. We echo this call and believe that Doubletree falls within the scope of the trade-off between probing load and the information gleaned from such probing. In this section, we discuss a brand new infrastructures that could take advantage of Doubletree and, thus, traceroute@home.

OneLab [21] is a European project, due to start in September 2006, that assembles some of the most highly respected network research teams from university and industry labs in Europe. OneLab aims to extend PlanetLab into new environments beyond the traditional wired internet, to deepen PlanetLab's monitor capabilities, and to provide a European administration for PlanetLab nodes in Europe.

OneLab's monitoring component is mainly motivated by the fact that many applications, such as those that take advantage of multihoming, could benefit from a better vision of the characteristics of the underlying network. Some objectives of the monitoring component are designing and implementing a prototype measurement infrastructure providing router and AS-level path information. The project also intends to submit the definition of a standard API for the measurement platform to the IETF (or IRTF).

This infrastructure has the potential to perform large-scale measurements. That is why we believe that Doubletree and, by extension, traceroute@home would perfectly fit into these plans. Further, we designed our implementation while keeping in mind extensibility. Changes or extensions needed when incorporating our prototype within the active measurement monitoring component of OneLab will be easy to achieve. In addition, due to the use of XML, the prototype is easy to tune and the resulting topological information might be quickly changed in an other format than XML if needed.

6   Conclusion

In this paper, we were interested in large-scale topology discovery at IP level. We focused on Doubletree, an efficient and cooperative topology discovery algorithm.

We put Doubletree one step further than its initial description by proposing a Java implementation. The application that implements Doubletree is called traceroute@home, is easily extensible, is open-source and freely available.

We discussed our implementation by explaining our design choices and by presenting the global functioning of the system. We next evaluated the performance of traceroute@home and described its behavior in a real environment. We finally discussed a monitoring infrastructure that could benefit of traceroute@home.

traceroute@home is an on-going project. We aim to improve our tool. In the near future, we would like to develop IPv6 libraries in order to permit IPv6 networks probing. Further, we are currently developing a peer-to-peer (or an overlay) system for managing the probing monitors and the entire structure. This new architecture is based on the prototype discussed in this paper.

Acknowledgements

Mr. Donnet's work was supported by a SATIN European Doctoral Research Foundation grant and by an internship at CAIDA.

References

1. Huffaker, B., Plummer, D., Moore, D., claffy, k.: Topology discovery by active probing. In: Proc. Symposium on Applications and the Internet. (2002)
2. Luckie, M.: IPv6 scamper (2005) WAND Network Research Group.
3. Georgatos, F., Gruber, F., Karrenberg, D., Santcroos, M., Susanj, A., Uijterwaal, H., Wilhelm, R.: Providing active measurements as a regular service for ISPs. In: Proc. Passive and Active Measurement (PAM) Workshop. (2001)
4. McGregor, A., Braun, H.W., Brown, J.: The NLANR network analysis infrastructure. IEEE Communications Magazine **38** (2000)
5. Lakhina, A., Byers, J., Crovella, M., Xie, P.: Sampling biases in IP topology measurements. In: Proc. IEEE INFOCOM. (2003)

6. Clauset, A., Moore, C.: Traceroute sampling makes random graphs appear to have power law degree distributions. cond-mat 0312674, arXiv (2004)
7. Cheswick, B., Burch, H., Branigan, S.: Mapping and visualizing the internet. In: Proc. USENIX Annual Technical Conference. (2000)
8. Anderson, D.P., Cobb, J., Korpela, E., Lebofsky, M., Werthimer, D.: SETI@home: An experiment in public-resource computing. Communications of the ACM **45** (2002)
9. Shavitt, Y., Shir, E.: DIMES: Let the internet measure itself. ACM SIGCOMM Computer Communication Review **35** (2005)
10. Donnet, B., Raoult, P., Friedman, T., Crovella, M.: Efficient algorithms for large-scale topology discovery. In: Proc. ACM SIGMETRICS. (2005)
11. Donnet, B.: traceroute@home 1.0 (2006) See `http://trhome.sourceforge.net`.
12. Donnet, B., Huffaker, B., Friedman, T., claffy, k.: Implementation and deployment of a distributed network topology discovery algorithm. cs.NI 0603062, arXiv (2006)
13. PlanetLab Consortium: PlanetLab project (2002) See `http://www.planet-lab.org`.
14. Donnet, B., Huffaker, B., Friedman, T., claffy, k.: Increasing the coverage of a cooperative internet topology discovery algorithm (2006) Under review.
15. Microsystems, J.S.: JDK 1.4.2 (1994) `http://java.sun.com`.
16. JSocket Wrench: Release R04 (2004) See `http://jswrench.sourceforge.net/`.
17. Peterson, L., Pai, V., Spring, N., Bavier, A.: Using PlanetLab for network research: Myths, realities, and best practices. Design Note PDN–05–028, PlanetLab Consortium (2005)
18. Donnet, B., Friedman, T., Crovella, M.: Improved algorithms for network topology discovery. In: Proc. Passive and Active Measurement (PAM) Workshop. (2005)
19. Planet-Lab users mailing-list: Very high ping/traceroute latencies on planet-lab nodes (2006) `http://lists.planet-lab.org/pipermail/users/2006-March/001892.html`.
20. claffy, k., Crovella, M., Friedman, T., Shannon, C., Spring, N.: Community-oriented network measurement infrastructure (CONMI) (2005) Workshop Report. Available from `http://www.caida.org/publications/papers/2005/conmi`.
21. OneLab consortium headed by Université Pierre & Marie Curie: The onelab project (2006) See `http://www.one-lab.org`.

The Virtual Topology Service: A Mechanism for QoS-Enabled Interdomain Routing

Fábio Verdi[1], Maurício Magalhães[1], Edmundo Madeira[2], and Annikki Welin[3]

[1] Department of Computer Engineering and Industrial Automation (DCA)
School of Electrical and Computer Engineering (FEEC)
State University of Campinas (Unicamp)
13083-970, Campinas-SP, Brazil
{verdi, mauricio}@dca.fee.unicamp.br
[2] Institute of Computing (IC)
State University of Campinas (Unicamp)
13084-971, Campinas-SP, Brazil
edmundo@ic.unicamp.br
[3] Ericsson Research-Sweden
annikki.welin@ericsson.com

Abstract. In this paper we present the Virtual Topology Service (VTS), a new approach to provide interdomain services taking into account QoS and Traffic Engineering (TE) constraints. It is known that in these days the provisioning of end-to-end interdomain connections does not consider any type of QoS due to limitations of the BGP routing protocol. At the same time, many extensions have been proposed to BGP, however none of them were put into practice. We advocate in favor of a service layer that offers new mechanisms for interdomain routing without affecting the underlying Internet infrastructure. The VTS abstracts the physical network details of each Autonomous System (AS) and is totally integrated with BGP. We use the Internet hierarchy to obtain more alternative routes towards a destination. The architecture was already used to provide interdomain services in optical networks. In this paper we show how the architecture can be used to provide interdomain connections in IP networks. We will show how the VTS and other services such as the end-to-end negotiation service work together to provide a complete mechanism for provisioning of interdomain QoS-enabled routes. Preliminary evaluation results are also presented.

1 Introduction

For a long time the networking research community has been trying to tackle with the interdomain provisioning of services. QoS is not considered in the original Internet architecture. Only a best-effort packet delivery service is available, but there is value in enhancing the network to meet application requirements [1]. While there are several solutions for TE and QoS within a single domain, the provisioning of services involving more than one domain is still a challenge. Currently, there is no way for end-users to choose the route in the domain-level.

G. Parr, D. Malone, and M. Ó Foghlú (Eds.): IPOM 2006, LNCS 4268, pp. 205–217, 2006.

Today, the decision of what route the packets of a given flow will follow is taken basically by the BGP protocol which considers the business relationships between each pair of domain. Due to this scenario, the interdomain selection of routes does not take into account the diversity of paths among domains as a solution for load balancing and traffic engineering.

Although the BGP is the current "de facto" interdomain routing protocol, it is becoming the main drawback for Internet Service Providers (ISPs). The BGP advertises only reachability among domains by announcing network prefixes to its neighboring domains. While protecting internal details of domains is a requirement in commercial relationships, a certain degree of information could be opened without affecting the local strategy of each domain. Such information might include an abstract cost representing the current physical state of the network. Another known problem of BGP is related to slow convergence behavior. Interdomain routes can take up to 15 minutes to fail-over in the worst case [2]. This is not acceptable for mission-critical applications.

In this work we present a proposal for interdomain provisioning of QoS-enabled routing in IP networks. We assume that every single domain is capable of offering QoS towards some destination network prefixes. Each domain is responsible for implementing the network-enabled QoS by using, for example, DiffServ or Multiprotocol Label Switching (MPLS) technologies. Each domain is represented by a Virtual Topology (VT) that gathers the traffic parameters to cross the domain in terms of bandwidth, latency, jitter, etc. The Virtual Topology Service (VTS) is responsible for getting the VT of each domain in a route in order to give the source domain more information related to QoS towards a given destination. The End-to-End Negotiation Service (E2ENS) will then negotiate the contracts with the chosen domains to reserve resources.

We elaborate our architecture to work in the management plane acting as a service layer for other domains and customers. This service layer offers specific services such as e2e interdomain provisioning of connections and interdomain provisioning of Virtual Private Networks (VPNs). The management plane abstracts the underlying details on how the provisioning of connections is performed by each network provider. This idea allows to have a service layer (a.k.a. service provider) over the network provider. In this work we propose to implement the service layer by using the Web services technology [3]. The main objective of Web services is to help organizations drive their business towards a service-oriented enterprise (SOE) [4].

The architecture presented in this paper has already been used for provisioning of interdomain services in optical networks. In [5] the architecture and the services are detailed. In [6] the evaluation of the architecture in terms of time and bandwidth consumption to establish interdomain optical connections was done. In this current paper, we detail how the architecture can be easily used to provide interdomain routing with QoS in IP networks.

Our approach does not preclude the Internet as it is today neither does it exclude the BGP policies that define the rules on how the network traffic must enter and exit in each domain. On the contrary, we propose a service layer that facilitates the interactions between providers by using Web services keeping all the legacy Internet infrastructure. Instead of being competitive with BGP, our architecture can be seen as a complementary tool for BGP running on the management plane.

This paper is organized as follows: next Section shortly presents some related works and their limitations for interdomain QoS routing. Section 3 details the VTS. Section 4 is dedicated to show how the VTS was implemented and integrated with BGP routing. Finally, Section 5 concludes the paper.

2   Related Works and Current Limitations of Interdomain QoS Routing

The most recent approach related to interdomain QoS routing is the MESCAL approach [7]. The MESCAL project has defined the concept of local classes and extended classes. Extended classes are created by combining local classes with other extended classes from external domains. After defining and engineering the classes, the architecture has a function that advertises the QoS capabilities to customers and peers. The authors claim that a variety of advertising mechanisms can be used. However, they do not discuss how the classes will be announced. They assume an abstract relationship called *peering* that is general and implies the existence of some type of customer-provider interaction.

Although the project idea is very interesting, it depends on the extension of the BGP, what, in our point of view, is a long term process of standardization without guarantees of becoming a standard. The authors advocate in favor of a QoS-enhanced BGP (q-bgp) that will be used to convey QoS-related information between ASes. Since the QoS information is exchanged by the q-BGP, the route selection process needs to be modified to take into account new QoS attributes. This is, to the best of our knowledge, difficult to be put into practice due to the changing that will be necessary in every border router across the Internet. Are the companies, ISPs and vendors willing to change their interdomain process of route selection? These challenges have limited the solutions that depends on the BGP extensions.

Among other works that propose to extend BGP, we can cite [8,9] that suggest an additional attribute called QOS_NLRI as an extension of the BGP UPDATE message. The work presented in [9] has proposed statistical QoS metrics to achieve satisfactory routing optimality. However, none of these proposals was put into practice in real scenarios.

3   The Virtual Topology Service

The Virtual Topology (VT) concept represents the QoS features of each domain towards a destination domain. A given domain may have several different VTs that are advertised following specific rules such as the variation of the amount of traffic during the day, services being offered (VoIP, video conference, VPNs being established and so forth) and the availability of the resources in terms of bandwidth, latency, jitter and packet loss rate. The VT is formed by a set of virtual links that map the current link state of the domain without showing internal details of the physical network topology. Fig. 1 illustrates the domain 1's VT and its QoS values towards the downstream domain 2.

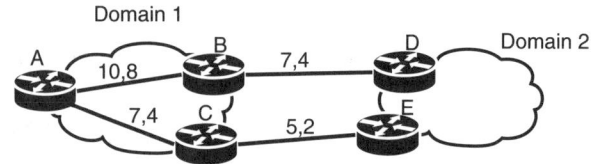

Fig. 1. The Virtual Topology concept

In this example, domain 1 has two egress routers towards domain 2. Each internal virtual link as well as each external virtual link can be implemented in different ways. One egress router can be a DiffServ router whilst the other can be an MPLS router. Each virtual link gathers QoS traffic parameters that reflect the current state of the domain. In Fig. 1, the virtual link A-B has, for example, 10 Mbs of available bandwidth and a latency of 8 ms. The virtual link A-C has 7 Mbs of available bandwidth and a latency of 4 ms. Then, going through the virtual link A-B to reach domain 2, the available bandwidth would be 7 Mbs (the lowest value is used) having a latency of 12 ms. If the virtual link A-C is chosen, the available bandwidth would be 5 Mbs having a latency of 6 ms. If Domain 1 advertises its virtual topology to Domain 2 at this moment, Domain 2 would see the virtual topology and the values as shown in Fig. 1.

The way the IP traffic will be carried in each domain depends solely on the engineering policies deployed within the domain. In case of hard QoS guarantees, specific MPLS LSP tunnels could be used within each domain to force the reservation of the resources otherwise DiffServ with TE functions should ensure the QoS required in the contracts.

3.1   How VTs Are Obtained

We have defined two models to obtain VTs: the push model and the pull model. The latter is also known as on demand model.

The Push Model: In this model, the VT advertising is done between pairs of domains. Each domain announces the VT to all its neighbors respecting commercial and economic relationships previously defined. The push model is more indicated to a regional scenario. We envisage that this regional scenario is formed by "condominiums of domains" by which a group of domains agrees on advertising virtual topologies to each other. This advertising is done in a peering model where all the domains that make part of the same condominium have the virtual topologies of other domains. These condominiums of domains could define business rules in a tentative of creating new relationships that make the interactions more customer-oriented.

The Pull Model: In the pull model, domains do not advertise their topologies to the neighboring domains. The VT is obtained by each domain that wants to know the VTs of other domains. When a given AS needs to find an e2e QoS-enabled interdomain route, it queries its BGP local table and verifies what are the possible routes to reach the destination. Then, based on these routes, the source domain can invoke each domain in the route towards the destination and gets the VT of the domains specifically for that route. Fig. 2 shows an example.

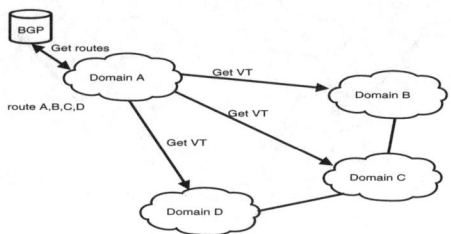

Fig. 2. Getting VTs (Pull Model)

Suppose that domain A needs to reach a destination located at domain D with a certain QoS. Domain A queries its local BGP table and discovers that the best route based on BGP is through domains B, C and D. Then, domain A invokes the VTS to obtain the virtual topology of each domain. Unlike the MESCAL approach that adopts a cascaded model to perform the interactions among domains, the VTS adopts a centralized mechanism by which all the VTs are obtained in parallel. Note that there can be more than one BGP path to the destination D. As a result, the source domain can recursively query each domain in each path and find the best path towards the destination.

After obtaining all the VTs of each possible route towards domain D, the source domain A is able to use a Constraint Shortest Path (CSP) algorithm to find the best route that fits the QoS requirements. The path calculation can be done using only one attribute or more than one. For instance, if a given domain requires low latency, it can use only the latency attribute. If it desires lowering packets loss rate, it can use only the packet loss rate attribute.

However, after obtaining the VTs of each route towards a destination, the source domain can realize that there is no route that satisfies the QoS requirements. Then, as mentioned before, our architecture makes use of the Internet hierarchy to collect more alternative routes towards a given domain. It does so by going one-level uphill in the Internet hierarchy to get other not-advertised BGP routes. This is explained below.

Although very difficult to define, the Internet hierarchy is divided into 5 layers [10]: the dense core(0) with about 20 ASes, transit core (1) with about 129 ASes, outer core (2) with 897 ASes, small region ISPs (3) with about 971 ASes and the customers (4) or stubs ASes with about 8898 ASes. Based on previous studies, it was verified that the quantity of possible different paths towards a given destination increases when going uphill in the Internet hierarchy [10]. As an example, there are 2409 edges from level 3 to level 4 and 3752 from level 2 to level 3. It has been also shown that there are about 193,000 different paths from any customer AS to the dense core [2]. However, the BGP only advertises the best path and then, the quantity of possible paths depends on the multi-homing aspect of each domain. Fig. 3 illustrates this scenario.

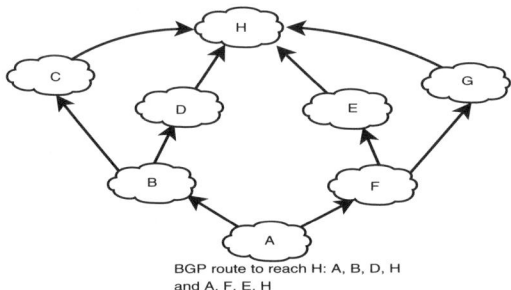

BGP route to reach H: A, B, D, H
and A, F, E, H

Fig. 3. Getting more paths (Pull Model)

The stub domain A is multi-homed with domains B and F. It has received two BGP routes from its providers to reach prefixes at domain H. The first route is A, B, D, H and the second one is A, F, E, H. However, there are other possibilities to reach H through domains C and G that were not advertised to domain A. Then, to increase the number of paths to query for VTs, the VTS can invoke its providers and asks the other BGP routes that were not advertised. In this case, domain B would return to domain A the path C, H towards H and domain F would return G, H towards H. As a consequence, domain A would have now other two different paths to reach H and then can ask each domain to get the VTs.

Due to the current characteristics of the Internet, going uphill only one level from a given domain is enough to have a high number of possible different paths. As mentioned in [10], as we move from customers to the core, the inter-level connectivity raises significantly. At the same time, not all the BGP routes need to be returned to the source domain. In a real Internet scenario with thousands of routes, only some of them should be selected based on local policies defined by the provider.

3.2   Detailing the Architecture

The architecture being proposed in this work offers a service layer by which QoS routing can be obtained among IP domains without having to change or extend the BGP routing. The architecture and the interactions between the modules are shown in Fig. 4. The service layer is formed by the Virtual Topology Service (VTS) and by the End-to-End Negotiation Service (E2ENS). The VTS is responsible for interacting with other domains to obtain the virtual topologies. The E2ENS is responsible for doing the negotiation among domains in order to establish e2e interdomain contracts. Since the focus of this paper is on the VTS, specific details of the E2ENS and the way these contracts are established in terms of format are not defined here and are left for further studies.

During the negotiation, the required QoS parameters are transferred from the head-end domain[1] to other domains in order to negotiate and establish a SLA between domains. We adopted a two-phase-star-based model by which the head-end domain negotiates with other domains to define a contract. The first phase queries the downstream domains about the possibility of reserving resources (basically bandwidth) for a given IP flow. During the first phase, the traffic parameters are analyzed in each downstream domain in order to verify if the IP flow can be accepted. The second phase confirms the contract with each domain considering that all the domains involved in the negotiation have agreed in receiving the IP flow.

In the first phase, the DiffServ Code Point (DSCP) or the MPLS label of the downstream domains should be returned to the head-end domain to configure the egress routers (in the second phase) of the upstream domains to swap data packets. This is necessary to mark the packets in order for them to be identified by the adjacent downstream domain as belonging to a specific QoS class.

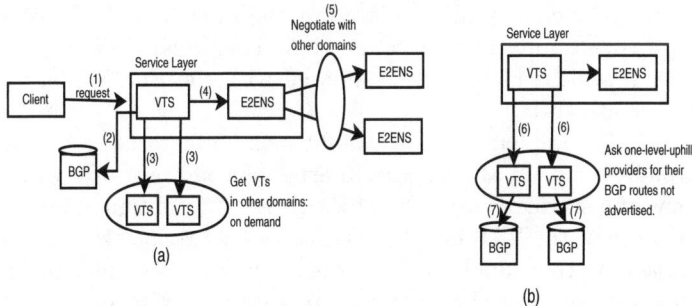

Fig. 4. Service Layer Architecture and the Interactions between the modules

Our approach works as follows. First of all, the source domain tries to attend the customer request by going through the local BGP routes towards the destination (Fig. 4 (a)). If such routes do not satisfy the customer requirements in terms of QoS, then the source domain can ask its providers for other BGP routes that were not advertised (Fig. 4 (b)). When a request coming from a client is done (step 1 in Fig. 4 (a)), the service layer validates such request and reads the BGP routing table (step 2 in Fig. 4 (a)) to obtain the routes available to reach the destination required by the client. After obtaining the BGP routes, the VTS is invoked to get the VTs of each domain belonging to the available routes (step 3 in Fig. 4 (a)). When all the VTs are gathered, the source domain will have a topological view in terms of QoS routing towards the destination. The source domain can then apply a CSPF algorithm and find the shortest path that satisfies the QoS requirements. After having calculated the interdomain path, the E2ENS will be invoked to perform the

---

[1] The head-end domain is the domain where the request was made.

negotiation of the traffic parameters between the domains and possibly establish a SLA (steps 4 and 5 in Fig. 4 (a)). If after negotiating with other domains, the local domain concludes that the VTs obtained taking into account the local BGP routes do not attend the customer requirements, then steps 6 and 7 in Fig. 4 (b) are executed. Step 6 asks the providers of the local domain for the BGP routes towards the destination that were not advertised by the BGP. After this phase, the local domain will have other routes and then steps 3, 4 and 5 in Fig. 4 (a) are again executed to get the VTs of the new routes and negotiate with the domains. If after executing these phases the local domain was not able of finding a route that satisfies the customer requirements, the request can be refused or sent using the best-effort forwarding.

4   Implementation and Validation

4.1   Implementation

The implementation of our architecture considers only the Pull Model. The Pull Model can be put into practice in a shorter period of time since it reflects a very practical scenario and considers the integration with the BGP routing. The Pull Model can then be incrementally replaced by the Push Model. The Push Model was used to provide interdomain connections in optical networks [6]. In this current paper it will be used to be compared with the Pull Model considering the IP network scenario.

In this work, the implementation validates the integration between the service layer (mainly the VTS) with the Internet routing protocol, i.e., the BGP. We show how the service layer interacts with BGP to obtain routes towards a destination and how the VTS interacts with other VTSs in other domains to get BGP routes and virtual topologies. Details about the Web Services infrastructure including aspects related to service registration, service look up and service binding are not considered in this paper. Such details can be found in [5,6].

To validate our approach, we deployed the architecture in a real scenario running BGP with eight domains. Each domain has its virtual topology represented in XML files. These XML files store the current state of the domain in terms of QoS. They are usually fed by the administrator (to define QoS classes) and by dynamic tools using probing mechanisms such as *ping* and *traceroute*. The border routers are running BGP daemons to exchange reachability information. Each border node is represented by a virtual machine. We have used the QEMU virtual machine [11] for this testbed. The BGP MIBs are obtained by using the UCD-SNMP suite [12] (currently called Net-SNMP). The communication between the BGP daemon and the UCD-SNMP agent was done by using the SMUX protocol. Fig. 5 shows the architecture of a node and how the integration between the service layer and the BGP is done (the VTS is responsible for interacting with the SNMP in each node. Then, only the VTS is shown in the figure). We created a gateway responsible for receiving the socket request from the service layer and converting it to a *SNMP get command*. The Web services were created using the Apache AXIS 1.2 [13]. The communication between Web

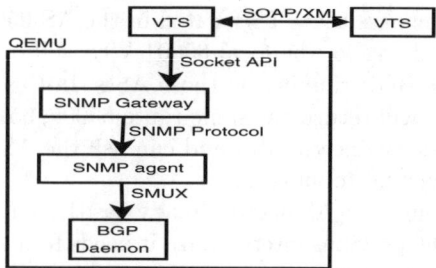

Fig. 5. Node Architecture and the Integration between the service layer and the BGP

services is done through XML-based Simple Object Access Protocol (SOAP) messages over HTTP. The wrapped SOAP binding was used as the model for Web services interactions.

4.2 Validation

Fig. 6 shows the scenario, the interdomain topology and the virtual topology of each domain used for this work. For sake of simplicity, each virtual link has only an abstract cost that represents the QoS of the virtual link.

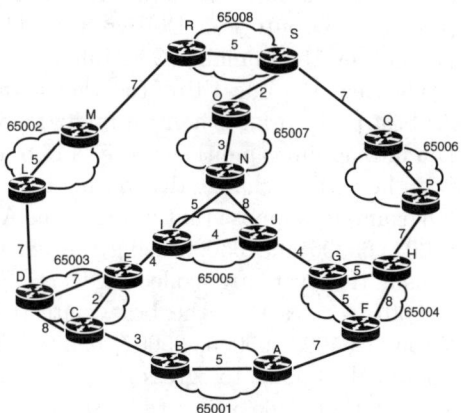

Fig. 6. Scenario used in our tests

In this scenario, the AS 65001 needs to send an interdomain QoS-enabled IP flow towards AS 65008. Firstly, AS 65001 gets its local BGP routes towards 65008. By observing its BGP routes it realizes that there are two paths to AS 65008: the path going through 65003, 65002, 65008 and the path going through 65004, 65006, 65008. Then, the VTs of each domain in both routes are obtained by invoking the VTS. After getting the VTs, the local domain calculates the best path and invokes the E2ENS to negotiate the traffic parameters in each domain. If the acceptance of the IP flow is not possible through either of these two routes, then the local VTS invokes the VTS located in the AS 65003 and VTS located in the AS 65004. By doing so, the local 65001 VTS will obtain the BGP routes not advertised by the BGP running on those ASes. In this case, both ASes (AS 65003 and AS 65004) will return the same route 65005, 65007, 65008. Then, the local AS 65001 knows two new routes and can ask the AS 65004, 65003, 65005 and 65007 for their virtual topologies.

Observe that although the AS 65001 already has the VTs of AS 65003 and AS 65004 (obtained in the previous interaction), it needs to invoke again those ASes to get the VTs. This is necessary because the VTS in each domain only returns the VT related to the route being analyzed. This is done in every domain. Such mechanism allows each domain to have a very fine control of its virtual topologies. Also, it avoids that the requesting domain gets not allowed information about other routes towards the destination.

After the source domain finds a route that satisfies the QoS requirements and the negotiation has established the contracts with every downstream domain, it is necessary to verify if the route chosen considering QoS constraints is the same as the BGP route chosen by the BGP protocol. Observe that before finding a QoS route, the data packets are being sent towards a given prefix by using the BGP route selected by the BGP algorithm. Then, if the QoS route is different from the one being used, i.e, the egress router and the next hop are different, the Local_Pref attribute of the BGP needs to be changed in order to modify the internal routing to use the egress router that represents the QoS-enabled route. Observe that the QoS-enabled route is a BGP route and as such it respects business relationships between the domains. The difference is that without the service layer QoS routing view, the Local_Pref together with other BGP metrics are used to choose the best path towards a given prefix following business rules. When the QoS is taken into account, another route can be selected and then the Local_Pref attribute can be used to change the routing in each domain following QoS constraints. As an example, suppose that in Fig. 6 the AS 65001 has selected the route through 65004, 65006 and 65008 using the normal traditional BGP route selection mechanism. However, after collecting the QoS-related information using the VTS, AS 65001 decides that the best route that satisfies the QoS requirements goes through 65003, 65005, 65007, 65008. Then, the Local_Pref attribute should be modified to point to egress router C.

We have evaluated our prototype using the scenario presented above. We analyzed the times comparing the Push Model with the Pull Model. The main difference between both is that in the Push Model, when a domain wants to send IP packets with QoS towards a destination, the virtual topologies would be present in the source domain because they were advertised earlier. In this case, only the path calculation and the negotiation are necessary. In the Pull Model (on demand model), the virtual topologies need to be obtained at the time of the request. We run 100 requests and collected the average time. The average time for the Push Model is 205 ms to attend each request. This includes the path calculation and the two-phase negotiation protocol. The Pull Model took 1 second on average considering only the local BGP routes. This is the time to read the BGP routes using the UCD-SNMP agent, collect the virtual topologies in each domain for each route, apply the CSP algorithm and negotiate with each downstream domain. When the Pull Model needs to go uphill one level in the hierarchy of our scenario, the average time to attend each request increased to 3 seconds. In our case, the VTS needs to obtain the BGP routes from AS 65003 and AS 65004. It is important to say that in the Pull Model, the time to invoke the SNMP agent going through our gateway as shown in Fig. 5 is 664 ms. This includes the communication time to invoke the SNMP gateway (Socket API), to invoke the SNMP agent and to parse the answer from the SNMP agent to return to the service layer. The command being used to read the BGP MIB is the *snmpwalk*. The SNMP agent performance depends on the size of the BGP table. In a real Internet scenario with thousands of entries, the *snmpbulkwalk* should be used.

4.3   Final Discussion

The contracts between domains should be established considering aggregated traffic demands so that flows to the same destination are seen as a single flow. This can be done by having a traffic matrix to estimate the amount of traffic towards the same destination. This aggregation avoids route oscillation and instability in the routing table. If every single IP flow were treated individually, the Local_Pref attribute should be changed every time a new route to the same destination is found to attend the new QoS requirements.

Dynamic IP flows should be aggregated into an already established QoS class. If not possible, the flow should be sent using the normal best-effort forwarding. However, there could be a mechanism for a given local domain to reroute all or some of the current IP flows in order to give QoS to new traffic demands. The local domain (through the VTS) could ask its downstream domains to rearrange the current IP flows without affecting the pre-established SLAs. If this rearranging is possible, then dynamic IP flows could be aggregated into a QoS class while keeping the QoS of all the previous flows. This issue is left for further study. Also, in more dynamic scenarios the Push Model should take into account a threshold to re-advertise the virtual topologies. Only when the threshold is reached (e.g. bandwidth lower than a given quantity), a new virtual topology is advertised.

The evaluation proved that the Push Model presents lower times than the Pull Model. However, the Push Model is more indicated to a regional scenario considering condominiums of domains. The Pull Model represents a more practical scenario and can be used in the Internet in a shorter period of time since the Push Model depends on how to group the condominiums. Our intention was not to evaluate all the scenarios but to prove the feasibility of our architecture in terms of integration with the Internet routing protocol. As can be seen, the service layer offers more information to the source domain. It can have a general view about the e2e QoS-enabled interdomain routing alternatives without changing the BGP engine.

5   Conclusion

In this paper we presented an architecture to provide a service layer over the current Internet infrastructure. This architecture aims at supporting new services for provisioning of interdomain QoS-enabled routing. Due to the limitations of BGP routing, there is a great effort to develop and deploy new mechanisms to offer new services in the Internet. Our architecture is based on the Virtual Topology Service that is responsible for offering routing information related to QoS among domains. Also, the negotiation service was used to negotiate traffic parameters, reserve resources and establish contract between domains. The use of the Web Services technology makes the architecture more business-oriented and facilitates the definition and the interaction of the services.

We analyzed the integration of the service layer with the BGP routing and proved that it is feasible in terms of how a source domain can start obtaining virtual topologies towards a given destination. The idea of going uphill in the Internet hierarchy is, to the best of our knowledge, a new way to obtain route alternatives for ASes. Based on previous studies, going one level uphill in the hierarchy is enough to have a high number of different routes towards the network prefixes.

We have previously used the virtual approach for interdomain optical networks and proved its feasibility. In this paper we migrated the architecture to attend IP networks and analyzed its advantages over the traditional Internet routing.

Acknowledgments

The authors would like to thank CAPES, CNPq, FAPESP, and Ericsson Brazil for their support.

References

1. NSF Report. Overcoming Barriers to Disruptive Innovation in Networking. Report of NSF Workshop, 2005.
2. S. Agarwal, C. Chuah, and R. Katz. OPCA: Robust Interdomain Policy Routing and Traffic Control. *OPENARCH*, 2003.
3. Web Services Activity: http://www.w3.org/2002/ws/.
4. F. L. Verdi, E. Madeira, and M. Magalhães. Web Services and SOA as Facilitators for ISPs. *International Conference on Telecommunications (ICT'06), Madeira Island, Portugal*, May 2006.
5. F. L. Verdi, R. Duarte, F. C. de Lacerda, E. Madeira, E. Cardozo, and M. Magalhães. Provisioning and Management of Inter-Domain Connections in Optical Networks: A Service Oriented Architecture-based Approach. *IEEE/IFIP Network Operations and Management Symposium (NOMS'06)*, 2006.
6. F. L. Verdi, E. Madeira, and M. Magalhães. On the Performance of Interdomain Provisioning of Connections in Optical Networks using Web Services. *IEEE International Symposium on Computers and Communications (ISCC'06), Sardinia, Italy*, June 2006.

7. M. P. Howarth et al. Provisioning for Interdomain Quality of Service: the MESCAL Approach. *IEEE Communications Magazine*, 43(6):129–137, June 2005.
8. G. Cristallo and C. Jacquenet. An Approach to Inter-domain Traffic Engineering. *Proceedings of XVIII World Telecommunications Congress*, 2002.
9. L. Xiao et al. Advertising Interdomain Qos Routing. *IEEE Journal on Selected Areas in Communications. Vol. 22. No. 10*, pages 1949–1964, December 2004.
10. L. Subramanian, S. Agarwal, J. Rexford, and R. Katz. Characterizing the Internet Hierarchy from Multiple Vantage Points. *IEEE Infocom*, June 2002.
11. http://fabrice.bellard.free.fr/qemu/, April 2006.
12. http://net-snmp.sourceforge.net/, April 2006.
13. http://ws.apache.org/axis/, 2005.

Correlating User Perception and Measurable Network Properties: Experimenting with QoE

Pál Varga, Gergely Kún, Gábor Sey, István Moldován, and Péter Gelencsér

[1] Budapest University of Technology and Economics,
Department of Telecommunications and Media Informatics, Magyar Tudósok krt. 2.,
H-1117 Budapest, Hungary
{pvarga, kun, moldovan, gelencser}@tmit.bme.hu
[2] AITIA International Inc., Czetz J. u. 48-50., H-1037 Budapest, Hungary
gsey@aitia.ai

Abstract. User perception of a networking service is usually very different from the operators' understanding of service usability. Quality of Experience (QoE) metrics are supposed to describe the service from the end-users' point of view – although QoE is hard to measure for mass services. Collection and analysis of QoS and SLS (Service Level Specification) properties of networking services are daily tasks of the operators. These metrics, however often provide misleading description of user satisfaction. Our ultimate aim is to find methods and metrics determining QoE by passive measurements on an aggregated network link. In this paper we describe our experimental results on correlating the severity of a network bottleneck and the experienced service quality. During our measurements we have loaded the network with various kinds of service requests and made notes on the perceived quality. We have also captured packet level traffic, and derived metrics based on packet inter-arrival times, packet size information and packet loss information. This paper briefly presents some of our analysis results.

1   Causes of Quality of Experience Degradation

Quality of Experience is measured – or more appropriately: perceived – at the end terminal, by the user. QoE shows similarities to Quality of Service (QoS), it does, however, differ from QoS by various means. QoE is subjective in nature, furthermore, service availability and user access capabilities get more focus in QoE than in QoS.

There can be several scenarios where the experienced service quality becomes less than satisfactory. The roughest QoE degradation is the unavailability of a service. This actually happens more regularly than one might expect. Examples of temporal and/or regular service-unavailability include an unsuccessful paying procedure at a webshop, a broken download, or a "webpage unreachable" message. These cases are very hard to measure from a central site, since continuous application monitoring and data processing is not always feasible. Another type of QoE violation is tied to network QoS, hence it can be correlated with some

G. Parr, D. Malone, and M. Ó Foghlú (Eds.): IPOM 2006, LNCS 4268, pp. 218–221, 2006.
© Springer-Verlag Berlin Heidelberg 2006

kind of QoS metrics. Network related QoE-degradation can manifest itself as extended response time, decreased download speed, less enjoyable audio and video streaming. This is usually caused by one or more network bottlenecks, where packets get congested, queued, then delayed or dropped.

There are only a few QoE-QoS measurements published at the moment. An example of these is [1], where the authors describe the effects of end-to-end delay of live video streaming on QoE. In this paper we describe our results on correlating the severity of a *network bottleneck* and the experienced service quality. There can be several link properties applied to characterize the presence of bottlenecks in packet data networks [2]. In the following section we describe our experimental results on correlating perceived QoE and properties of some bottleneck metrics.

2   Measurement Scenario, Methods and Results

Our aim is to correlate the end-user experience with measurable properties of the traffic. The metrics that correlate better with the end-user experience will be more effective QoE measures than the ones that do not show direct correlation. To reach this aim we have set up a configurable DSL-like environment in a segment of our university department, where the department staff was the suffering object of our QoE testing. Altogether about 50 end-user machines were involved in the test. We have degraded the usual 100 Mbps Internet access to 1 Mbps DSL-like lines. Moreover, we have introduced an artificial bottleneck for the department segment. Later this bottleneck was changed from 7 Mbps to 2.5 Mbps in six steps.

To obtain more subjective results, the downgrade was not linear; sometimes we have upgraded the service - pretending as if it was going to be normal. We have allowed each of these bottleneck scenarios to last 30 minutes. During the tests we have captured the traffic at two measurement points, and made notes on our experiences as network application users. The measurement scenario is depicted by Figure 1.

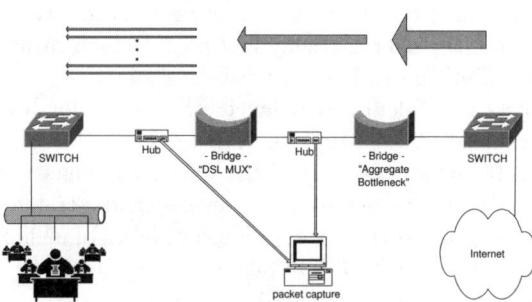

Fig. 1. Measurement architecture with the relations of download volumes

We have extensively used various networking applications and services during the test. Applications included audio and video streaming, web-browsing, ftp and peer-to-peer downloads, Internet-telephony (Skype), database access and many more. Table 1 shows our experiences in function of the bottleneck-severity.

Table 1. Perceived quality of network services at different bottleneck scenarios

Available Bandwidth	Video stream	Audio stream	Skype	P2P traffic	Web browsing
7.0 Mbps	perfect	perfect	perfect	33-95 kBps	good
5.0 Mbps	occasional squares in picture	perfect	perfect	33-90 kBps	good
4.5 Mbps	playing pauses at key-frame changes	perfect	perfect	33-88 kBps	slow
4.0 Mbps	sec-long pauses at key-frame changes	perfect	perfect	12-14 kBps	slow
3.5 Mbps	bad; squares and pauses are regular	short pauses	scratching	3-10 kBps	unbearably slow
2.5 Mbps	very bad; not enjoyable	longer pauses	scratching	3 kBps	extremely bad

To correlate QoS with QoE, we have derived packet-level properties from the collected traces. As a basic analysis, these included throughput, packet retransmission (TCP packet loss). As a more advanced analysis, we have calculated the delay factor of flows [2] in order to understand what the users may experience as application delay. Furthermore, we have derived the metric called PIT kurtosis (the 4th central moment of Packet Interarrival Time distribution) [3]. This measure indicates the severity of queuing in the traffic path.

Figure 2 introduces our results by presenting the most interesting of them, thus from the seven scenarios only two could be seen: the 7 and the 4.5 Mbps cases. When the available bandwidth (ABW) was 7 Mbps the network was not overloaded: all the applications worked properly, while at 4.5 Mbps the network load caused service degradations in case of some applications (see Table 1).

Figure 2.a presents four average metrics calculated for the two scenarios. We have found (like other authors [4]) that the calculation of an average packet loss ratio cannot reflect the severity of the congestions. Likewise the throughput metric: the measured amount of traffic was approximately the same in the two cases. In spite of these the average delay factor and much rather the kurtosis of PIT PDF shows a significant difference between them.

Figure 2.b depicts the calculated delay factor values for 3 minutes long intervals for both scenarios. At 7 Mbps there is no severe congestion, however at 4.5 Mbps, there is. In the second part of the 4.5 Mbps scenario the delay factor diminished, mainly due to the decrease of user-generated traffic. Certainly users gave up trying to access network services after a number of unsuccessful requests. This also explains that the difference of *average* delay factors is smaller than the PIT PDF kurtosis values' for the 7 and 4.5 Mbps cases (see Figure 2.a).

Figure 2.c and d present the number of resent packets in function of time, calculated for each 10 ms. Considering the two graphs there is no significant difference between them. TCP packet loss type metrics and retransmission counts

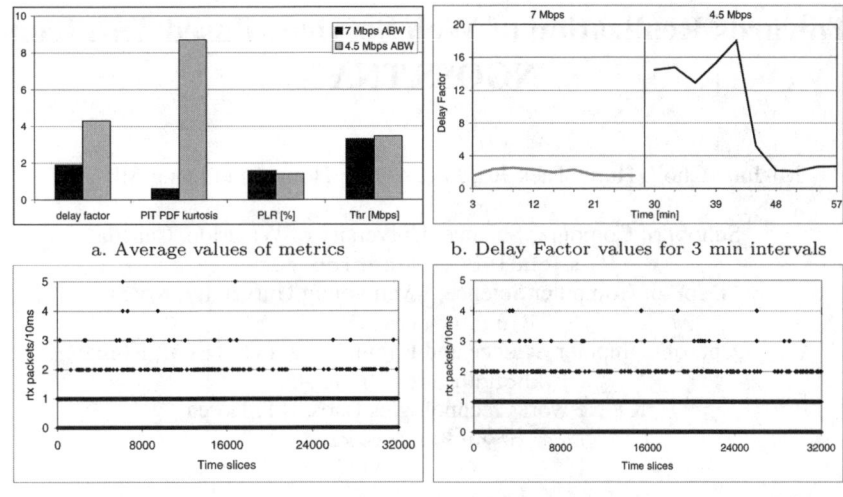

a. Average values of metrics   b. Delay Factor values for 3 min intervals

c. Number of retransmission at 7 Mbps   d. Number of retransmission at 4.5 Mbps

Fig. 2. Measurement results for two available bandwidth scenarios

may suggest some network misbehavior only in very special cases, but in general these measures do not provide really useful information, even if it is analyzed for short time periods.

3 Summary

This paper investigated network performance metrics and aimed to correlate these with the end-users' Quality of Experience. Our standpoint is that the metrics used for detecting bottlenecks are able to reflect QoE as well. In order to prove our ideas we have carried out a number of measurement scenarios in real network environment with different bottleneck conditions, some of which were presented in the paper. Presently we are working on further analysis of traffic behavior in bottleneck network situations, and in QoS aware networks.

References

1. Rahrer,T., Fiandra R., Wright,S., Triple-play Services Quality of Experience (QoE) Requirements and Mechanisms, DSL Forum Working Text WT-126, February 21, (2006)
2. Varga, P., Kún, G., Fodor, P., Bíró, J., Satoh, D., Ishibashi, K.: An advanced technique on bottleneck detection. In IFIP WG6.3 Workshop, EUNICE 2003, (2003)
3. Varga, P., Kún, G.: Utilizing Higher Order Statistics of Packet Interarrival Times for Bottleneck Detection. In IFIP/IEEE E2EMON05, (2005)
4. Vacirca, F., Ziegler, T., Hasenleithner, E.: An Algorithm to Detect TCP Spurious Timeouts and its Application to Operational UMTS/GPRS Networks. Journal of Computer Networks, Elsevier, (2006)

Towards Realization of Web Services-Based TSA from NGOSS TNA*

Mi-Jung Choi[1], Hong-Taek Ju[2], James W.K. Hong[3], and Dong-Sik Yun[4]

[1] School of Computer Science, University of Waterloo, Canada
mjchoi@cs.uwaterloo.ca
[2] Dept. of Computer Science, Kyemyoung University, Korea
juht@kmu.ac.kr
[3] Dept. of Computer Science and Engineering, POSTECH, Korea
jwkhong@postech.ac.kr
[4] KT Network Technologies Labs, KT, Korea
dsyun@kt.co.kr

Abstract. To avoid frequent changes of OSS's architecture, TNA provides NGOSS architecture in technology-neutral manner. TNA can be mapped to appropriate TSAs using specific technologies such as XML, Java and CORBA. Web Service can be applied for NGOSS TSA. In this paper, we examine architectural principles of TNA and propose an application mechanism of Web services technologies to TNA.

1 Introduction

As the data communications, telecommunication and other types of communications converge, the complexity and heterogeneity of networks and provided services are rapidly increasing. So, future Operation Support Systems (OSSs) need to cope with these incremental changes and complexity of networks and services. To meet this requirement, TeleManagement Forum (TMF) provides a Next Generation Operations Systems and Software (NGOSS) framework for rapid and flexible integration of operation and business support systems [1].

The network services and resources should be easily modified by appropriate business goals and policies. As new network services emerge, the operation management system also needs to be newly developed and use up-to-date technologies. To avoid frequent changes and upgrades of OSS's architecture, TMF provides NGOSS architecture in technology-neutral manner. Technology-neutral architecture (TNA) [1] is the basic concept and component of NGOSS architecture. This TNA can be mapped to appropriate technology-specific architectures (TSAs) and this mapping can leverage industrial standard frameworks such as service-oriented architecture (SOA), component-based architecture and distributed computing.

---

* This research was supported in part by the MIC (Ministry of Information and Communication), Korea, under the ITRC (Information Technology Research Center) support program supervised by the IITA (Institute of Information Technology Assessment) (IITA-2005-C1090-0501-0018) and by the Electrical and Computer Engineering Division at POSTECH under the BK21 program of the Ministry of Education, Korea.

G. Parr, D. Malone, and M. Ó Foghlú (Eds.): IPOM 2006, LNCS 4268, pp. 222–227, 2006.

Web service is one possible technology for NGOSS architecture. Web service is an emerging technology, which is being standardized continuously. Web service is also distributed and services-oriented computing technology with strong support from the industry. Therefore, the mapping of NGOSS TNA to Web services-based TSA is a promising research area.

The 'realization' in this paper means that how essential parts of the NGOSS architecture can be implemented using a specific standards-based COTS technology. We choose Web service as the specific technology. In this paper, we first examine NGOSS TNA in the perspective of the concepts, requirements and components. Then, we propose a technology specific architecture using Web services technology.

2 Architecture

In this section, we describe the NGOSS TNA and the architecture of the current KT's OSS [2]. The difference in the architecture is also examined.

2.1 NGOSS TNA

The insulation of system architecture from technology details provides a number of related benefits [5]: First, it ensures the validity of the NGOSS architecture over time by supporting the deployment of new technologies without having to re-architect the entire OSS solution. Second, it provides the architectural underpinnings for the simultaneous use of multiple technologies in a single integrated OSS environment, supporting legacy systems and enabling technology migration over time. Finally, insisting that the architecture remain technology neutral helps to prevent system

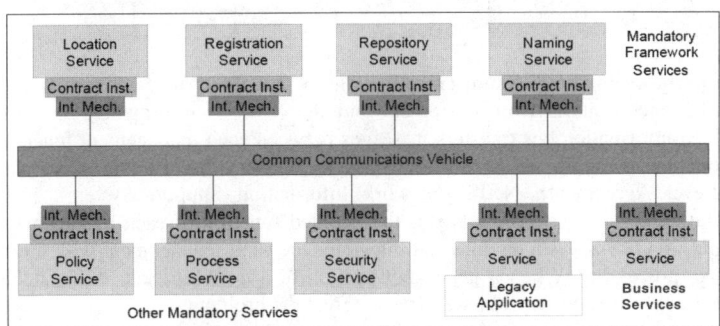

Fig. 1. This shows the detailed views of NGOSS technology-neutral architecture [1]. The service modules can communicate with each other through common communications vehicle (CCV) [1]. CCV is a kind of message bus independent of technology. The services are divided into mainly two parts: business services and framework services [3]. Business services provide the application level functionality that directly supports the implementation of a business process such as SLA management, billing mediation, QoS and etc. Framework services provide the infrastructure necessary to support the distributed nature of the NGOSS TNA. Contract [4] is the central concept of interoperability. Contract Instance is a runtime manifestation of an NGOSS contract implementation that provides one or more functions to other runtime entities. It represents the binding unit.

design decisions being taken too early in the development lifecycle, so the architecture is protected from over-specification. An excess of design detail in the core of the architecture would make it more difficult to find technologies with which to implement it.

2.2 KT NeOSS

Currently, KT has developed and used New Operation Support System (NeOSS) [2]. NeOSS is an integrated OSS platform for all KT services. But, NeOSS does not fully support NGOSS's architecture. NeOSS is more focused on NGOSS's business process framework such as enhanced telecom operations map (eTOM) and shared information data (SID).

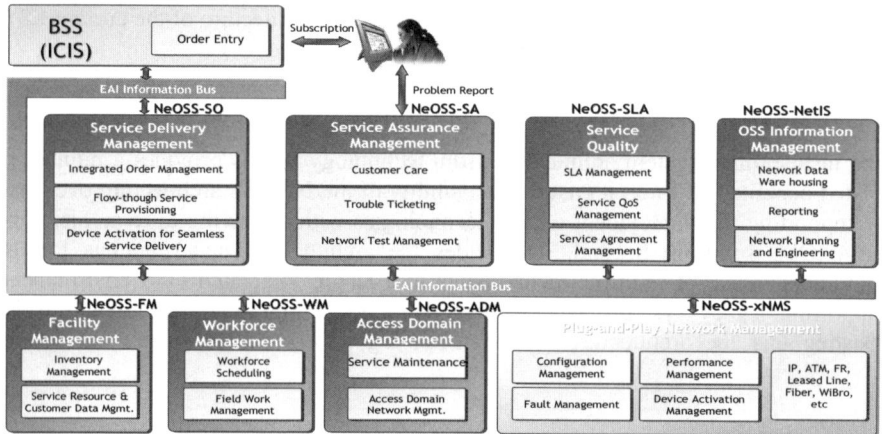

Fig. 2. This shows the architecture and functions of NeOSS. The BSS system receives the service order and requests from customers and delivers the information to correspondent modules. EAI information bus transfers messages between each management function module. NeOSS function modules are NeOSS-SO (Service Order), SA (Service Assurance), SLA (Service Level Agreement), NetIS (Network Information Support System), FM (Facility Management), and etc. The yellow box is the detailed function of each management module. For example, NeOSS-SA is a module that integrates the fault management with the concept of flow-through processing. When it gets a notification of a trouble ticket, it dispatches relative information to the affected services in order to correct the problem.

The difference between NGOSS TNA and NeOSS' architecture is as follows. EAI information bus of NeOSS plays a role of CCV of TNA. However, NeOSS does not have the concept of pre- and post-conditions. NeOSS does not have the concepts of maintaining invariant attributes and behavior, another crucial concepts of contracts. Therefore, there is an interface between management function modules in NeOSS. Moreover, NeOSS includes management services such as SLA, SA, SO, FM and etc. and these services provide management functions for the business process. But, NeOSS does not include framework services of NGOSS TNA, so it does not support the distributed nature of TNA.

3   Alignment of NGOSS TNA

In this section, we present a method to how the technology maps to the concepts of the NGOSS TNA and additionally describe detailed mappings of the technology-neutral shared information and contract models. The core architectural principles to be examined for alignment of NGOSS TNA and TSA are below.

First, the architecture needs to provide the distribution support. The architecture provides for communication between business services and between process control and other system services. This entails support for location independence, distributed transactions and other aspects of component/service interaction in a distributed systems environment. The architecture does not stipulate the precise mechanism for communication with services or any inter-service protocol, other than the technology-neutral methods for conveying business semantics and content between participating entities.

Second, the architecture should support the separation of business process from software implementation. The decoupling of business process from service implementation requires that the interface and semantics of services be rigorously defined in a technology-neutral way. This is achieved by the NGOSS "Contract"- a mechanism for formally defining the business semantics of a distributed service. A contract is initially defined in technology-neutral form, expressing the business functionality of the service. Technology specific information required for service invocation is added later, and includes such information as the supported communications protocols and invocation patterns. The NGOSS contract registry provides support for run-time discovery of services based on Contracts.

Finally, the OSS needs to provide the shared information for management information model. NGOSS stipulates that any business information that needs to be shared between services should be considered the property of the enterprise as a whole and not of any particular application or component. It uses the concept of shared information, which describes all shared business information in a specific NGOSS deployment. Information services coordinate access to information through well-defined interfaces and are responsible for information integrity and for managing information stewardship, where specific entities in the system have transient responsibility for a shared information entity. To design technology-specific architecture, the technology must align with these NGOSS architectural principles.

TMF proposed two technologies of XML [5] and CORBA [6] for NGOSS TNA and specified technology application notes with these technologies. OSS/J initiative proposed Java technology as a specific technology for TNA [7]. XML has a natural affinity with communications management. The use of XML to validate, manipulate and define the structure of application specific information models using general-purpose tools becomes an attractive possibility. However, XML has a substantial overhead associated with the text-only encoding of data. Also, XML/HTTP solutions (e.g. SOAP) suffer as yet from the lack of availability of common distributed processing support services provided by more mature platforms, e.g. CORBA and J2EE services [5].

CORBA is a distributed processing platform. Therefore, CORBA supports communication method and framework services for distributed processing. Also, CORBA supports interface definition with specifying CORBA IDLs. But, CORBA does not provide information modeling, so CORBA can define shared information using XML or other languages [6]. J2EE directly implements the principles of NGOSS TNA such as distribution support and separation of business process from implementation. But, J2EE architecture does not have any explicit support for the concept of shared information or federated information services as defined in NGOSS TNA [7].

4   Our Approach Using Web Services

Web services are very complex and include a lot of standard specifications. At this moment, the standardization efforts are still going on. So, it is not easy to specify the appropriate Web services technologies applicable to NGOSS architecture. We more focus on the NGOSS specification from a Web services perspective. Our to-be architecture is extended from this TNA architecture using Web services technologies.

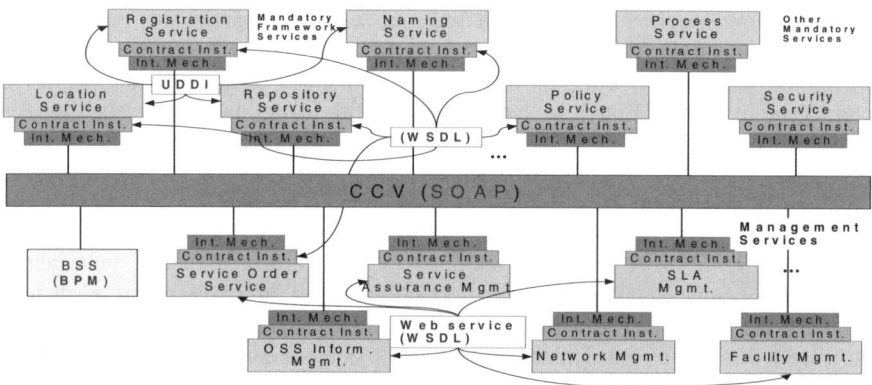

Fig. 3. This shows our proposed technology-specific architecture. XML is used for the definition and encoding of messages in data communication protocols. SOAP is a protocol that can be used as a CCV-based solution. Web Service Definition Language (WSDL) can be used to define contracts between process entities through SOAP. UDDI provides a comprehensive mechanism for locating services at run time by storing service interfaces in their WSDL format. That is, UDDI supports framework services. Application programs can search the UDDI repository to find the interface that they require, download the WSDL description and use the binding information to communicate with the interface over a suitable communication channel. Other framework services and management operation services such as SLA, NM and SA can be defined as new services using WSDL. Web Services technologies can be applied in a number of areas to assist with the management of business processes. Web Services Business Process Execution Language (WS-BPEL) can define business process and the BPEL engine handles the process sequentially.

5   Concluding Remarks

We examined the concept, architectural principles and components of NGOSS TNA. We proposed our technology-specific architecture using Web services. Our work is the early stage, so we need to concrete our TSA with more knowledge of Web services technology.

To validate our proposed architecture, we will implement a prototype of Web services-based TSA. Then, we will implement one of management operation services of NGOSS such as QoS management or SLA and perform a test on our implementation system. We will apply our proposed architecture to the architecture of KT's NeOSS. Finally, we will extract performance metrics of Web services-based TSA and conduct performance analysis.

References

1. TMF, "The NGOSS Technology-Neutral Architecture", TMF053, Version 5.3, Nov. 2005
2. KT, "NeOSS (New Operations Support System)", http://www.kt.co.kr
3. TMF, "NGOSS Architecture Technology Neutral Specification - Distributed Transparency Framework Services", TMF 053F, Version 1.0, Jan. 2004
4. TMF. "NGOSS Architecture Technology Neutral Specification - Contract Description: Business and System Views", TMF 053B, Version 4.4, Feb. 2004
5. TMF, "NGOSS Phase 1 Technology Application Note - XML", TMF057, Version 1.5, Dec. 2001
6. TMF, "NGOSS Phase 1 Technology Application Note - CORBA", TMF055, Version 1.5, Aug. 2001
7. C. Ashford, "OSS through Java as an Implementation of NGOSS", White paper, Apr. 2004

An Efficient Queue Management (EQM) Technique for Networks

Debessay Fesehaye Kassa

Department of TELIN
Ghent University
Gent, Belgium
debessay@telin.ugent.be

Abstract. In this paper I present a simple and efficient queue management (EQM) technique based on optimal bandwidth allocation for networks. Each router calculates the bandwidth share (throughput)[1] of each interfering flow at a link using link capacity and Interface IP Address (see RFC 3630) or round trip time (RTT) and congestion window size of the flows which are carried in the packets. The sources send packets and the routers in the path drop or allow the packets based on a certain probability and label the non-dropped packets with the allowable optimal throughput for the next round. Each router in the path modifies this label and the sources eventually adjust their sending rates based on the allowable throughput they get from the returning ACKs. In this way EQM finds the fair bandwidth allocation and gives fair queue management.

I also a prove that EQM can converge to a stable point.

1 Introduction

In the following sections I give a brief description of how EQM works, its convergence and a brief summary.

2 How EQM Works

Let $S_i, I_i, C_i, Q_i, d_i, L_i$ be the spare (unused) bandwidth, link interference (number of interfering flows in this case), link capacity, router queue size, control interval and total number of packets during the control interval at link i respectively. Now I set the throughput T_j of each flow j to be proportional to the total allowable bandwidth $C_i - Q_i/d_i$. That is

$$T_j \quad \propto \quad C_i - Q_i/d_i.$$

---

[1] These throughput values can be treated as link weights. A shortest path algorithm (Dijkstra) can be used to find the path with the highest allowable throughput. Instead of summing up the link weights as is the case in the normal Dijkstra algorithm, the highest throughput can be obtained by taking the mimimum of the allowable throughput in the path.

G. Parr, D. Malone, and M. Ó Foghlú (Eds.): IPOM 2006, LNCS 4268, pp. 228–231, 2006.
© Springer-Verlag Berlin Heidelberg 2006

But the sum of T_j of each interfering flow at link i should be $C_i - Q_i/d_i$. This implies that

$$T_j = \frac{C_i - Q_i/d_i}{I_i} = \frac{C_i d_i - Q_i}{d_i I_i} = \frac{C_i d_i - Q_i}{\sum_{j=1}^{L_i}(RTT_j/w_j)}$$

where I_i can also be estimated by the number of local IP addresses which are listed in the local interface IP address field of a link as shown in RFC 3630 [4]. Otherwise the optimal throughput T_j can be obtained using the RTT and congestion window sizes of the flows which are carried in the packets passing through the routers in a given control interval d_i.

Each router updates T_j at every control interval $d_i = \frac{\sum_{j=1}^{L_i} RTT_j}{L_i}$ where the round trip times RTT_j can be obtained from the timestamp option (see RFC 1323) by making small modification. The modification is that the two four-byte timestamp fields (*TSval* and *TSecr*) should contain the previous and current timestamp values of the sender from which any router in the path can get the round trip time of a flow to which the packet passing through it belongs. Details on the use and changes of the timestamp options with regard to different operating system implemetations will be discussed in the extended version of this paper. As an alternative other simpler estimates of the control interval can be used.

From the *cwnd* and RTT values in the arriving packets, the routers in the path also obtain the average throughput R_i in a control interval d_i. If $R_i > T_j$ router i drops a fraction $(R_i - T_j)/R_i$ of all L_i packets which arrive in a control interval d_i. In this case each of the arriving packet is dropped with probability $1 - T_j/R_i$. If $R_i \leq T_j$ packets arriving in a control interval d_i are forwarded. Each router in the path modifies the throughput label of the non-dropped packets as is the case in CSFQ [5]. Similar to CSFQ the average value R_i can also be replaced with the throughput value of each flow in each packet in the above calculation. A packet of flow j with a throughput R_j is then forwarded to the next hop with the probability $min(1, T_j/R_j$. To reduce per packet overhead and computation each packet can be made to carry the throughput of its corresponding flow which is $R_j = w_j/RTT_j$ instead of the congestion window. Then each router modifies this value according to the above rule. That is the new R_j is the minimum of R_i and the previous R_j.

When the ACK packets return to the source with the maximum allowable throughput in the path, the respective source j with RTT_j sets its *cwnd* to $T_j RTT_j$ and sends packets accordingly. At each arrival of an ACK the source can send w'_j/w_j where w'_j and w_j are the current and previous *cwnd* values respectively.

As another simple alternative EQM can also replace the random early detection (RED) protocol. Here instead of labeling the packets, router i can replace the *cwnd* value in the data and/or ACK packets with the T_j value. This technique may not require that data packets follow the same path discovered by the ACK packets. However the convergence of EQM can be slower as the feedback is received after about one RTT and the path of the packets may change.

3   Convergence Proof

Let us consider a network of N links. Let Λ_i denote the aggregate load offered to link i, $1 \leq i \leq N$. Let's assume that link i has I_i interfering flows. Let's define a bottleneck link to be a link from which a flow gets its smallest share of bandwidth in its path. With this definition each flow has only one bottleneck link. Let link i be the bottleneck link of flow i from which it gets the net traffic load of λ_{ii}. Let $h_i(\Lambda_i)$ be the maximum bandwidth share of flow i from its bottleneck link i. This value is bounded above by the link capacity minus the maximum queue rate the router of the link can accommodate. This is because EQM makes sure that each source doesn't send more than the link and router capacities. It must also be noted that $\lambda_{ii} \leq h_i(\Lambda_i)$ as some of the $h_i(\Lambda_i)$ packets sent by flow i may be dropped before they reach link i. If link i is not a bottleneck of flow j which has link j as a bottleneck and if flow j gets a maximum bandwidth share of $\hbar_j(\Lambda_i)$ from link i by EQM then we also have that $h_j(\Lambda_j) \leq \hbar_j(\Lambda_i)$.

Now we have that

$$\Lambda_i = \sum_{j=1}^{I_i} \lambda_{ji} = \lambda_{ii} + \sum_{j=1,\, j \neq i}^{I_i} \lambda_{ji}$$

$$\leq h_i(\Lambda_i) + \sum_{j=1,\, j \neq i}^{I_i} h_j(\Lambda_j) \leq h_i(\Lambda_i) + \sum_{j=1,\, j \neq i}^{I_i} \hbar_j(\Lambda_i)$$

$$\leq H(\Lambda_i) \tag{1}$$

for some function H.

As each of the $h_j(\Lambda_i)$ is bounded above by the capacity of link i minus the maximum queue rate the router of the link can accommodate, the function H is defined in a bounded polyhedron (region). Using extension of the Brouwer's fixed point theorem [1] for discontinuous functions given in [3], the function H has a fixed point. As is mostly the case[2] H can be shown to be a continuous decreasing function[3]. This is to say that if the total load in a given link increases the bandwidth shares of each flow decreases. Hence there exists a unique fixed point Λ_i^* such that $H(\Lambda_i^*) = \Lambda_i^*$ for $1 \leq i \leq N$. This is because of the fact that an increasing function ($f(\Lambda_i) = \Lambda_i$) and a continuous decreasing function like H or a continuous constant function meet at exactly one point.

4   Summary

A simple and efficient queue management (EQM) technique and a proof that it can converge to stability is given. EQM is expected to outperform existing queue management techniques such as CSFQ [5] and RED [2].

I am currently implementing EQM using simulation packages. I will further validate it using the fixed point theory. A simple extension of EQM can give quality of serivice (QoS) solutions.

---

[2] I am working on the details of EQM to see if there are points where this is not the case.
[3] If there are points where H doesn't satisfy these criteria, I will modify EQM to meet these conditions.

References

1. K. C. Border. *Fixed Point Theorems with Applications to Economics and Game Theory*. Press Syndicate of the University of Cambridge, The Pitt Building, Trumpington Street, Cambridge, United Kingdom, 1985.
2. S. Floyd and V. Jacobson. Random Early Detection gateways for Congestion Avoidance. *EEE/ACM Transactions on Networking*, 1(4):397–413, Aug. 1993.
3. J. J. Herings, G. van der Laan, D. Talman, and Z. Yang. A fixed point theorem for discontinuous functions. In *Tinbergen Institute Discussion Papers 05-004/1, Department of Econometrics and Tinbergen Institute*, Vrije Universiteit, De Boelelaan 1105, 1081 HV Amsterdam, The Netherlands, Dec. 2004.
4. D. Katz, K. Kompella, and D. Yeung. RFC 3630 - Traffic Engineering (TE) Extensions to OSPF Version 2. In *rfc3630.txt*, Berkeley, CA, USA, Sept. 2003.
5. I. Stoica, S. Shenker, and H. Zhang. Core-Stateless Fair Queueing: A Scalable Architecture to Approximate Fair Bandwidth Allocations in High Speed Networks. In *SIGCOMM*, Vancouver, British Columbia, CANADA, Sept. 1998.

Monitoring MIPv6 Traffic with IPFIX*

Youngseok Lee[1], Soonbyoung Choi[1], and Jaehwa Lee[2]

[1] Dept. of Computer Engineering, Chungnam National University,
220 Gungdong, Yusonggu, Daejon, Korea, 305-764
{lee, wakusoon}@cnu.ac.kr
[2] Korea Telecom, Woomyondong, Sochogu, Seoul, Korea
jhlee@noc.kr.apan.net

Abstract. As Mobile IPv6 (MIPv6) networks are being deployed, traffic measurement meets several challenges such as monitoring handover events, detecting tunneled IPv6 traffic, and classifying MIPv6 packets with the extension headers. However, typical traffic monitoring methods used in the plain IP network cannot be applied to MIPv6 networks. Hence, this paper proposes a new traffic monitoring mechanism suitable for a MIPv6 network. For this purpose, we used the IP Flow Information eXport (IPFIX) standard for monitoring MIPv6 access routers, and extended IPFIX templates that can carry MIPv6-specific information such as MIPv6 handover messages and IPv6-in-IPv6 tunneled flows. Thus, IPv6 data traffic, MIPv6 handover events, and tunneled IPv6 traffic could be monitored from multiple MIPv6 routers.

Keywords: MIPv6, handover, IPFIX, flow, traffic measurement.

1 Introduction

MIPv6 [1] protocols provide uninterruptible communication with mobile nodes. In MIPv6 networks, a care-of address will be associated with a mobile node (MN) whenever MN visits a foreign network without changing the MN's home address. Therefore, multiple IPv6 addresses will be available with MN. If MN moves to a foreign link, handover[1] will be carried out by exchanging Binding Update (BU) and Binding Acknowledgement (BA) messages with the home agent (HA). After registration of the BU message at HA, the traffic between MN and the correspondent node (CN) will be forwarded to the new IPv6 address of MN through tunneling via HA. If routing optimization is supported, the traffic from CN will be directly destined to the new address of MN.

---

* This research was supported by the MIC (Ministry of Information and Communication), Korea, under the ITRC (Information Technology Research Center) support program supervised by the IITA (Institute of Information Technology Assessment). (IITA-2005-(C1090-0502-0020)).

[1] In this paper, we assume only the bidirectional tunneling MIPv6 communication mode without route optimization. However, route optimization and other handover mechanisms such as fast handover could be easily supported.

G. Parr, D. Malone, and M. Ó Foghlú (Eds.): IPOM 2006, LNCS 4268, pp. 232–235, 2006.
© Springer-Verlag Berlin Heidelberg 2006

Herein, traffic measurement meets new challenges because of mobility and multiple addresses in MIPv6 networks. First, handover events should be monitored at every MIPv6 access routers. For instance, all the mobile nodes roaming over several foreign links in a large-scale MIPv6 network should be tracked for accounting and security. Second, a handover IPv6 data flow between CN and MN makes traffic measurement and analysis more complicated. For instance, after handover under no route optimization, MN associated with a new care-of IPv6 address will receive traffic from CN through the tunnel via HA. Therefore, the traffic monitoring method for MIPv6 networks has to be able to detect the dynamically tunneled traffic.

Traffic measurement at high-speed routers is generally carried out with the flow-level monitoring method, because packet-level measurement approach, that can generate correct results, is not easy to support fast line rates, and it is expensive for deployment and management. Hence, Internet Service Providers (ISPs) prefer routers or switches with built-in traffic monitoring functions. Recently, flow-level measurement at routers such as Cisco NetFlow [2] has become popular, because flow-level measurement could generate useful traffic statistics with a significantly small amount of measurement data. Based on Cisco NetFlow, the traffic monitoring mechanism at routers is being standardized by IETF IPFIX WG [3]. The key concept of IPFIX is the flexible and extensible template architecture that can be useful for various traffic monitoring applications such as IPv6 traffic monitoring, intrusion detection, and QoS measurement.

In this paper, we propose a MIPv6 traffic monitoring architecture based on IPFIX which can collect and analyze handover events, IPv6 flows, and tunneled IPv6 flows in MIPv6 networks. Therefore, we defined IPFIX templates for carrying MIPv6 BU/BA messages and tunneled IPv6-in-IPv6 traffic flows.

2 Traffic Measurement for MIPv6

2.1 IPFIX-Based MIPv6 Traffic Monitoring Architecture

As shown in Fig. 1, every MIPv6 access router has the IPFIX flow-level traffic monitoring functions of classifying packets into flows and exporting IPFIX-formatted flows to the collector. A central flow collector gathers IPFIX flows from all the MIPv6 access routers through reliable transport protocols such as Stream Control Transport Protocol (SCTP) or TCP[2]. It is assumed that each MIPv6 access router is synchronized with the stratum 1 time server in order to analyze flows captured at multiple routers. The IPFIX-based MIPv6 traffic monitoring process at routers will export IPFIX flows regarding MIPv6-specific information as well as basic IPv6 flow information. A flow is typically specified with 5-tuples of TCP/UDP/IP header fields: (src IP, dst IP, src port, dst port, protocol). The IPFIX-based MIPv6 traffic monitoring function at routers performs several flow-manipulating functions such as creating, updating, deleting, and exporting flow entries which include flow attributes such as 5-tuples, the number of bytes, packets, the first/last flow time, and MIPv6 information fields.

---

[2] Optionally, UDP may be used.

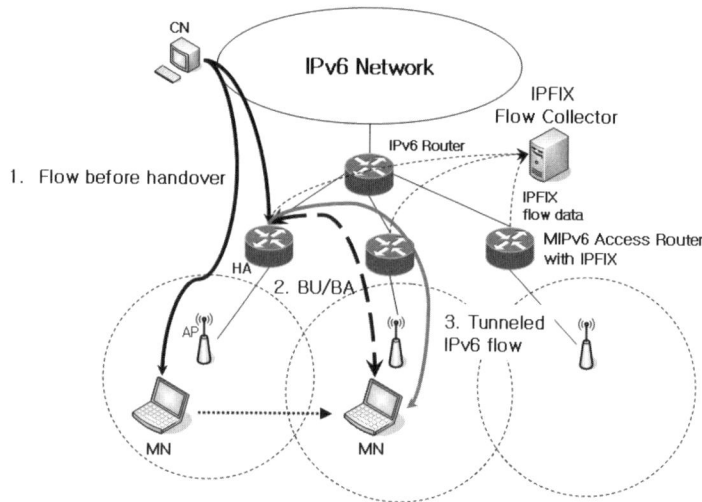

Fig. 1. IPFIX-based traffic measurement architecture for MIPv6 networks

2.2 IPFIX Template for Monitoring MIPv6 Traffic

In order to monitor MIPv6 traffic under IPFIX, we need appropriate IPFIX templates that can carry information regarding IPv6 flows, BU/BA messages, and tunneled IPv6 flows. The IPFIX template for IPv6 traffic has been defined and is being already used in Cisco NetFlow v9. Hence, in this paper, we define new IPFIX templates for BU/BA messages as well as IPv6-in-IPv6 tunneled flows.

When BU/BA IPv6 packets are exchanged between MN and HA during handover as shown in Fig.1, every packet should be investigated by routers to examine the cascaded IPv6 extension headers, because BU/BA messages are encapsulated with destination option/routing, ESP, and mobility headers in order in addition to the IPv6 basic header. After handover is completed, the traffic from CN to MN will be sent to MN through the tunnel via HA like Fig. 1. Therefore, the IPv6 extension headers for IPv6-in-IPv6 tunneled packets should be identified.

For these purposes, two new MIPv6-specific IPFIX templates shown in Fig. 2 are employed by IPFIX flow generators at access routers. Figure 2-(a) illustrates the IPFIX template format that could carry BU/BA messages. The BU/BA flow template consists of MIPv6messageType (= BU or BA), MIPv6CareOfAddress, MIPv6Home AgentAddress, and MIPv6HomeAddress besides the basic IPv6 template. This BU/BA flow is created when a BU or a BA packet has been monitored at MIPv6 routers. In Fig. 2-(b), the IPFIX template of IPv6-in-IPv6 tunneled flows is shown, where the source and destination addresses of the tunnel have been added to the basic IPv6 template.

After receiving IPFIX flows exported by MIPv6 routers, the IPFIX flow collector can detect the L3 handover events based on BU/BA flows and can track which flow as well as which MN have moved to which cell. In addition, the L3 handover latency can be derived from the timestamps of BU/BA flows. Moreover, the user-experienced

Version=1 0		Length = Total Length
Export Time		
Sequence Number		
Source ID		
Set ID = 2		Length = 60
Template ID = 257		Field Count = 5
0	Src IPv6 addr = 27	Field Length = 16
0	dst IPv6 addr = 28	Field Length = 16
0	Src port = 11	Field Length = 4
0	dst port = 11	Field Length = 4
0	Protocol = 4	Field Length = 4
0	ToS = 5	Field Length = 4
0	First time = 22	Field Length = 4
0	Last time = 21	Field Length = 4
0	OctetDeltaCount = 1	Field Length = 4
0	packetDeltaCount = 2	Field Length = 4
0	MIPv6messageType = 200	Field Length = 4
0	MIPv6CareOfAddress = 201	Field Length = 16
0	MIPv6HomeAgentAddress = 202	Field Length = 16
0	MIPv6HomeAddress = 203	Field Length = 16

(a) IPFIX template for BU/BA

Version=1 0		Length = Total Length
Export Time		
Sequence Number		
Source ID		
Set ID = 2		Length = 60
Template ID = 258		Field Count = 5
0	Src IPv6 addr = 27	Field Length = 16
0	dst IPv6 addr = 28	Field Length = 16
0	Src port = 11	Field Length = 4
0	dst port = 11	Field Length = 4
0	Protocol = 4	Field Length = 4
0	ToS = 5	Field Length = 4
0	First time = 22	Field Length = 4
0	Last time = 21	Field Length = 4
0	OctetDeltaCount = 1	Field Length = 4
0	packetDeltaCount = 2	Field Length = 4
0	IPv6TunnelSrcAddr = 300	Field Length = 16
0	IPv6TunnelDstAddr = 301	Field Length = 16
0	TunnelProto = 302	Field Length = 4

(b) IPFIX template for IPv6-in-IPv6 tunneled flow

Fig. 2. IPFIX templates for monitoring MIPv6 flows

handover latency could be calculated from the IPv6 data flows exported by the previous router with HA and the new router at the foreign network.

3 Conclusion

In this paper, we proposed a new traffic monitoring architecture for MIPv6 networks based on the IPFIX standard. Our traffic measurement approach can provide MIPv6-specific information in the mobile network such as handover events, and it can identify flows before/after handover. Though the proposed mechanism describes the basic MIPv6 handover, it could be easily applied to route optimization as well as to other fast handover schemes. Therefore, network management in a large-scale MIPv6 network will benefit from our traffic monitoring method with IPFIX.

References

[1] D. Johnson, C. Perkins, and J. Arkko, "Mobility Support in IPv6," IETF RFC3775, June 2004.
[2] Cisco NetFlow, http://www.cisco.com/warp/public/cc/pd/iosw/ioft/netflct/tech/napps_ipfix-charter.html
[3] J. Quittek, T. Zseby, B. Claise, and S. Zander, "Requirements for IP Flow Information Export (IPFIX)," IETF RFC3917, Oct. 2004.
[4] L. Deri, "nProbe: an Open Source NetFlow Probe for Gigabit networks," TERENA Networking Confernce, 2003.

Author Index

Printing: Mercedes-Druck, Berlin
Binding: Stein + Lehmann, Berlin

Lecture Notes in Computer Science

For information about Vols. 1–4180

please contact your bookseller or Springer